Exploratory Causal Analysis with Time Series Data

Synthesis Lectures on Data Mining and Knowledge Discovery

Editors

Jiawei Han, *University of Illinois at Urbana-Champaign*
Lise Getoor, *University of Maryland*
Wei Wang, *University of North Carolina, Chapel Hill*
Johannes Gehrke, *Cornell University*
Robert Grossman, *University of Chicago*

Synthesis Lectures on Data Mining and Knowledge Discovery is edited by Jiawei Han, Lise Getoor, Wei Wang, Johannes Gehrke, and Robert Grossman. The series publishes 50- to 150-page publications on topics pertaining to data mining, web mining, text mining, and knowledge discovery, including tutorials and case studies. Potential topics include: data mining algorithms, innovative data mining applications, data mining systems, mining text, web and semi-structured data, high performance and parallel/distributed data mining, data mining standards, data mining and knowledge discovery framework and process, data mining foundations, mining data streams and sensor data, mining multi-media data, mining social networks and graph data, mining spatial and temporal data, pre-processing and post-processing in data mining, robust and scalable statistical methods, security, privacy, and adversarial data mining, visual data mining, visual analytics, and data visualization.

Exploratory Causal Analysis with Time Series Data

James M. McCracken

ISBN: 978-3-031-00781-1   paperback
ISBN: 978-3-031-01909-8   ebook

DOI 10.1007/978-3-031-01909-8

A Publication in the Springer series
SYNTHESIS LECTURES ON DATA MINING AND KNOWLEDGE DISCOVERY

Lecture #12
Series Editors: Jiawei Han, *University of Illinois at Urbana-Champaign*
            Lise Getoor, *University of Maryland*
            Wei Wang, *University of North Carolina, Chapel Hill*
            Johannes Gehrke, *Cornell University*
            Robert Grossman, *University of Chicago*
Series ISSN
Print 2151-0067   Electronic 2151-0075

Exploratory Causal Analysis with Time Series Data

James M. McCracken

ISBN: 978-3-031-00781-1   paperback
ISBN: 978-3-031-01909-8   ebook

DOI 10.1007/978-3-031-01909-8

A Publication in the Springer series
SYNTHESIS LECTURES ON DATA MINING AND KNOWLEDGE DISCOVERY

Lecture #12

Series Editors: Jiawei Han, *University of Illinois at Urbana-Champaign*
              Lise Getoor, *University of Maryland*
              Wei Wang, *University of North Carolina, Chapel Hill*
              Johannes Gehrke, *Cornell University*
              Robert Grossman, *University of Chicago*

Series ISSN
Print 2151-0067   Electronic 2151-0075

Exploratory Causal Analysis with Time Series Data

James M. McCracken
George Mason University

SYNTHESIS LECTURES ON DATA MINING AND KNOWLEDGE DISCOVERY #12

ABSTRACT

Many scientific disciplines rely on observational data of systems for which it is difficult (or impossible) to implement controlled experiments. Data analysis techniques are required for identifying causal information and relationships directly from such observational data. This need has led to the development of many different time series causality approaches and tools including transfer entropy, convergent cross-mapping (CCM), and Granger causality statistics.

A practicing analyst can explore the literature to find many proposals for identifying drivers and causal connections in time series data sets. Exploratory causal analysis (ECA) provides a framework for exploring potential causal structures in time series data sets and is characterized by a myopic goal to determine which data series from a given set of series might be seen as the primary driver. In this work, ECA is used on several synthetic and empirical data sets, and it is found that all of the tested time series causality tools agree with each other (and intuitive notions of causality) for many simple systems but can provide conflicting causal inferences for more complicated systems. It is proposed that such disagreements between different time series causality tools during ECA might provide deeper insight into the data than could be found otherwise.

KEYWORDS

time series causality, leaning, exploratory causal analysis

To Angel, whose patience I have tested often but never broken

Contents

Preface

Consider a scientist wishing to find the driving relationships among a collection of time series data. The scientist probably has a particular problem in mind, e.g., comparing the potential driving effects of different space weather parameters, but a quick search of the data analysis literature would reveal that this problem is found in many different fields. They would find proposals for different approaches, most of which are justified with philosophical arguments about definitions of causality and are only applicable to specific types of data. However, the scientist would not find any consensus of which tools consistently provide intuitive causal inferences for specific types of systems. The literature seems to lack straightforward guidance for drawing causal inferences from time series data. Many of the proposed approaches are tested on a small number of data sets, usually generated from complex dynamics, and most authors do not discuss how their techniques might be used as part of a general causal analysis.

This work was developed from the realization that drawing causal inferences from time series data is subtle. The study of causality in data sets has a long history, so the first step is to develop a loose taxonomy of the field to help frame the specific types of approaches an analyst may be seeking (e.g., time series causality). Then, the philosophical causality studies must be carefully and deliberately divorced from the data causality studies, which is done here with the introduction of *exploratory causal analysis* (ECA). Finally, examples need to be presented where the different approaches are compared on identical data sets that have strongly intuitive driving relationships. Using such an approach, the analyst can develop an understanding of how a causal analysis might be performed, and how the results of that analysis can be interpreted. This work presents all three of these steps and is intended as an introduction and guide to such analysis.

James M. McCracken
March 2016

Acknowledgments

This work would not be have been possible without the help of many people who have supported me over the last few years, financially and otherwise. I have, hopefully, made those people aware of how much I appreciate their help. Most importantly though, I'd like to thank my adviser Professor Robert Weigel. I appreciate that he was willing to let me prove that I was a serious student, despite my somewhat unusual circumstances as a doctoral candidate. His academically holistic advising style emphasized a deep understanding of both the background literature and the many different methods for approaching modern data analysis problems. I believe he has helped me learn not only the material printed in this lecture but, more importantly, how to be a professional physicist. I could not have had a better adviser.

James M. McCracken
March 2016

CHAPTER 1

Introduction

Data analysis is a fundamental part of physics. The theory-experiment cycle common to most sciences [1] invariably involves an analyst trying to draw conclusions from a collection of data points. Gerald van Belle outlines data analysis as four questions in what he calls the "Four Questions Rule of Thumb" [2] (distilled from Fisher [3]), "Any statistical treatment must address the following questions:

1. What is the question?

2. Can it be measured?

3. Where, when, and how will you get the data?

4. What do you think the data are telling you?"

These items are each addressed by one of the three primary components of data analysis,

- the question (i.e., what is the purpose of the data analysis?; van Belle item # 1),

- the data (i.e., what measurements are available to address the question?; van Belle items # 2 and 3), and

- the tools (i.e., what methods, techniques, or approaches can be used on the data to address the question?; van Belle item # 4).

This work focuses on the tools (the third component) given data in a specific form (the second component) and a distinct, limited question (the first component). The concern of this work is data analysis to answer the question, "Does a given pair of time series data potentially contain some causal structure?" In physics, this question is common when data has been collected over time (i.e., time series data) in an experiment that was not intentionally designed to verify some existing theory, e.g., recording daily precipitation and temperature at a specific location for many years. The physicist's preliminary analysis of such data would be *exploratory* [4], and a part of that exploratory analysis may involve investigations of potential causal structure. The word "potential" is important to the analysis, as confirmation of any causal structure in the system would require further analysis, possibly including the collection of more data and/or the design on new experiments to collect new data. Such analysis would be *confirmatory* and one of the main purposes of the exploratory analysis is to guide confirmatory analysis [4].

Causal inference as a part of data analysis (referred to in this work as *data causality*; see Section 2) is a topic of debate among many researchers and has been for many years (see [5] for an introduction). In 1980, Clive Granger stated

> "Attitudes toward causality differ widely, from the defeatist one that it is impossible to define causality, let alone test for it, to the populist viewpoint that everyone has their own personal definition and so it is unlikely that a generally acceptable definition exists. It is clearly a topic in which individual tastes predominate, and it would be improper to try to force research workers to accept a definition with which they feel uneasy. My own experience is that, unlike art, causality is a concept whose definition people know what they do not like but few know what they do like." [6]

Granger was referring to the attitudes of statisticians, economists, and philosophers of the time who often repeated the platitude "correlation is not causation" [7] and declared statements of causality were only possible in experiments with randomized trials, as described by Fisher [3, 8]. General statements of causality in modern social science data analysis often still require Fisher randomization in the experimental design [9, 10]. It has come to be recognized, however, that randomized trials can be prohibitively difficult (e.g., too expensive) or impossible to conduct in practice (e.g., subjecting one subject to multiple treatments simultaneously) [9, 10]. Techniques have been developed for causal inference when Fisher randomization is unavailable. It has been argued that, in principle, Fisher randomization is not possible in any social science[1] experiment and thus modern data causality techniques are required for rigorous causal inference of such data sets [11].

Physics has a long history of studying causality, often more closely related to philosophical considerations rather than specific data sets of physical systems (see, e.g., [12]). Physics applies the scientific tradition of a prediction-experiment-repeat cycle [1] to fundamental systems, such as billiard balls and particle interactions, where randomization of hidden variables is usually not much of a concern. Some physicists consider statements of causality to be impossible without direct intervention into the system dynamics [13]. Modern data collection techniques, however, include measurements of system dynamics for which interventions are not technologically feasible. One such data set discussed in this work in is NASA/GSFC's Space Physics Data Facility's OMNIWeb (or CDAWeb or ftp) service, and OMNI data set [14]. Causality studies of such systems require data causality tools rather than traditional experiments (i.e., interventions into the system dynamics).

1.1   NOTATION AND TERMINOLOGY

Notation and terminology in causality studies can be contentious (see, e.g., [5, 15, 16]). This issue is discussed in detail in Sections 1.2 and 1.3. The notational shorthands "→" and "←" are used

---

[1]"Social science" is used here to mean any field of study of sufficiently complex systems, including sociology, psychology, economics, and the medical sciences.

to replace causal verbs such "cause" or "drive," e.g., $A \leftarrow B$ means B drives A. This shorthand is convenient but should not be conflated with causal terminology outside the scope of such terminology used in this work, i.e., $A \rightarrow B$ means A causes B using the definition of "cause" outlined in Sections 1.2 and 1.3. The word "tool" is also used extensively throughout this work to mean a technique, method, approach, or specific mathematical calculation (or any collection of such things) used as part of a data analysis. This terminology is, again, used for convenience. Causality studies have an extensive history and is broadly interdisciplinary. The use of nondescript, overarching terms like "tool" and notations such as "$\rightarrow$" are intended to prevent the discussion from getting lost in the details of nomenclature. The hope is that the notation and terminology used in this work do not obscure any underlying concepts or ideas. Such issues will be discussed whenever there may be confusion.

1.2   TIME SERIES CAUSALITY FRAMEWORK

Time series causality is causal inference made using only time series data. It is a term that may be applied to a wide array of different methods and techniques, as discussed in Section 3. It may be considered a subset of data causality, which is causal inference based on data of some kind. The specific type of data used in time series causality, i.e., time series, is the defining feature. A rough taxonomy of causality studies is presented in Section 2.

There is some debate in the literature about whether or not time series causality tools should be considered *causal* rather than simply *statistical* [15, 17, 18]. Most of the arguments against using such tools for causal analysis are equivalent to one of the three following truisms (or platitudes, depending on the author's point of view),

1. Correlation is not causation

2. Confounding needs to be addressed

3. *Proving* causation requires intervention into the system dynamics (i.e., experiments or "manipulation" [16]).

The first point may be argued as the motivation for some time series causality tools (see, e.g., [7]) but is often pointed out as a limitation of others (see, e.g., [19]). Truisms 2 and 3, however, have been argued as the primary reason times series causality tools should not be considered causal at all.

For example, Pearl [15] defines a *statistical parameter* as "any quantity that is defined in terms of a joint probability distribution of observed variables, making no assumption whatsoever regarding the existence or nonexistence of unobserved variables" and a *causal parameter* as "any quantity that is defined in terms of a causal model ... and is not a statistical parameter." He goes on to say, "Temporal precedence among variables may furnish some information about ... causal relationships ... [and] may therefore be regarded as borderline cases between statistical and causal models. We shall nevertheless classify those models as statistical because the great majority of policy-related questions *cannot* be discerned from such distributions, given our commitment to making no assumption regarding the presence or absence of unmeasured variables. Consequently, ... 'Granger causality' and 'strong exogeneity' will be classified as statistical rather than causal" (emphasis in original). Most times series causality tools rely on the temporal structure of the data sets for causal inference, and most draw conclusions only from the available data. Thus, following Pearl's definitions, most times series causality tools are statistical tools, not causal tools.

Holland [16] argues that any method or technique that does not involve intervention into the system dynamics may only be considered associational, not causal. Holland does not restrict the concept of interventions to experimenters conducting carefully designed experiments; he considers, e.g., the possibility of drawing causal inferences from observational data. He does, however, restrict the notion of an intervention by restricting the notion of a cause, which may not be an attribute of system (e.g., race or gender in social studies) and must be applicable equally to all units within the system. See [16] for a full discussion of these ideas. Most time series causality tools, as mentioned previously, draw causal inference from the data alone. There is usually no effort made to determine if one time series fits any specific definition of "cause" with respect to the other time series being analyzed. Thus, following Holland's arguments, such tools are associational tools, not causal.

Time series causality tools are used in this work for *exploratory causal analysis*, which is explained in more detail in Section 1.3. Labeling the causal inference as "exploratory" may, cynically, be seen as escamotage meant to avoid issues such as those raised by Pearl and Holland. It may be argued, however, that the three points listed above, along with issues of labeling and taxonomy, are of little importance to the task at hand. There are situations in which an analyst will have nothing but a set of times series. Should the analyst refuse to make any causal inference at all? Doing so would ignore previous scenarios in which analysts have successfully applied techniques, such as times series causality tools, to draw causal inferences that were later confirmed either by collecting more data and applying other data causality tools (such as those advocated by Pearl [15]) or by conducting experiments. The modifier *exploratory* applied to these causal inferences implies the need for further investigation and/or theory to fully understand the causal structure of the system. An analyst need not limit their data analysis toolbox by ignoring time series causality tools simply because such tools may only be "associational" or "statistical." Indeed, the inferences drawn with such tools may not be causal, but they could be, so they should not be ignored during a search for causal inference.

1.3   EXPLORATORY CAUSAL ANALYSIS FRAMEWORK

Exploratory causal analysis uses the term *exploratory* similar to the way it is used by Tukey in his approach to exploratory data analysis [4]. The preface of Tukey's original work provides general themes to both the exploratory data analysis he was advocating and exploratory causal analysis, including emphasis that the main idea is "... about looking at the data to see what it seems to say." and the primary concern is "... with appearance, not with confirmation" [4]. Exploratory analysis is not confirmatory analysis, and this point must be especially clear to a causal inference analyst. Tukey makes this idea clear with a bullet point in the Preface, "leave most interpretations of results to those who are experts in the subject-matter field involved" [4]. Times series causality tools are statistical tools, and statistical tools are necessarily only associational [5]. The relationships found with such tools can only be deemed "causal" with the application of outside theories or assumptions [5], which requires an analyst seeking to make such general causal statements to be well-versed in any such theories involving the analyzed data. Often the analyst is seeking to *confirm* a given causal theory. Statistical tools, such as time series causality tools, can certainly be used for such a task but not in the manner they are applied in this work. Such analysis would be *confirmatory causal analysis*, and this work is about *exploratory causal analysis*.

The distinction between exploration and confirmation in causal analysis is important because the analysis can (and should) be approached differently. Analysis meant to confirm some existing causal theory will be attempting to find specific patterns within the data. Presumably, there will be known (or assumed) models (or sets of models) to which the data will be fit. In such analysis, a primary goal might be understanding the residuals from the fits and special attention may be given to the sampling procedure to ensure the causal inferences are applicable to the desired populations. Exploratory causal analysis is not concerned with specific models² or the applicability of causal inferences beyond the data being analyzed. Exploratory causal analysis is characterized by a myopic goal to determine which data series from a given set of series might be seen as the primary driver. The exploratory causal inference is never assumed to apply to any data other than that being analyzed and is meant to guide the confirmatory causal analysis, especially in situations where existing outside theories or assumptions about the data are incomplete. The results of exploratory causal analysis may guide the design of new experiments, i.e., the collection of new data sets, from which new exploratory causal inferences may be drawn. The results may also guide the creation of new theoretical models or the development of new assumptions, which would become part of the confirmatory causal analysis. In short, exploratory causal analysis is meant to be the first step of causal inference. It is meant to draw as much potentially causal information as possible from the given data.

Another important reason for the distinction between exploration and confirmation in causal analysis is philosophical. Exploratory causal inference is intended to determine if (and what) causal structure may be present in a time series pair but is not intended to *confirm* such structure. There are scenarios in which any time series causality tool may incorrectly assign causal structure or may incorrectly not assign causal structure [5]. Furthermore, proof of causal relationships is often considered impossible with data alone [5, 9, 15]. Care must be taken to ensure that causal terminology used during exploratory causal analysis, such as "driving," "cause," "effect," and even "causality," is not conflated with other definitions of those terms. If, in an exploratory causal analysis of a given set of data series, one is identified as the "cause" of another, then this

---

²This is not meant to imply that exploratory causal analysis does not use models, only that such models are not assumed to reveal any underlying truths about the system dynamics or populations from which the data has been sampled. See Section 3.1 for a discussion of data models in exploratory causal inference.

identification should not be construed as a statement that the former is a *true* cause of the latter, in any (empirical, physical, metaphysical, philosophical, etc.) sense of the word "true." The identification of a "cause" among a given set of data series during exploratory data analysis is a statement that, assuming one of the data series may be interpreted as driving the others, the "cause" series is driving the "effect" series, where the technical meaning of "driving" depends on the specific causal inference tools being used but, in general, implies some data or data structure (i.e., the "cause") in the past of one series predicts, or raises the probability of, some data or data structure (i.e., the "effect") in the present or future of another series. For example, if a cause is identified as **A** in a pair of series $(\mathbf{A}, \mathbf{B})$ using transfer entropy,[3] then the exploratory causal statement **A** causes (or drives) **B** (written as $\mathbf{A} \rightarrow \mathbf{B}$) is the statement that the information flow[4] from **A** to **B** is greater than the information flow from **B** to **A**. It is a statement about a comparison of magnitudes of specific calculations conducted on the data series. The only causal interpretation given to the statement is that which the tool itself possesses (which, in this case, is one of "information flow" being indicative of causal relationships). No attempts are made to relate the causal interpretations of the tools used during exploratory causal analysis to more general notions of causality, e.g., in physics and philosophy. Such tasks would be part of any confirmatory causal inferences made with these tools.

Exploratory causal analysis is defined more by an approach to the analysis than the use of any specific set of tools or assumptions. As Tukey notes, exploratory analysis is "detective in character"[5] [4]. However, the specific exploratory causal analysis techniques used in this work rely on time series causality tools, which, in turn, rely on two important assumptions. The first is that the cause must precede the effect. This assumption has been argued among philosophers for a long time (see, e.g., [5, 15, 20–23]) and has also been questioned among physicists [12, 13]. The second assumption is that a driver may be present in the set of data series being analyzed. This assumption makes the exploratory causal inference susceptible to confounding [5, 15]; i.e., exploratory causal analysis is not intended to detect potential driving from sources outside of the data series set being analyzed. For example, the exploratory causal inference may be $\mathbf{A} \rightarrow \mathbf{B}$ when, in fact, there is some unknown data series **C** which may be more appropriately labeled the driver of both, i.e., $\mathbf{C} \rightarrow \mathbf{A}$ and $\mathbf{C} \rightarrow \mathbf{B}$. Any data-driven causal inferences are susceptible to confounding [5]. However, exploratory causal analysis, as it is done in this work, may be seen as particularly vulnerable given the desire to draw conclusions from only the time series data under consideration. Neither of these assumptions will be defended beyond noting that the agreements between intuitive causal "truths" and exploratory causal inferences shown in Section 4 imply these assumptions are valid in at least some circumstances.

---

[3]See Section 3.2.

[4]"Information flow" is defined in Section 3.2.

[5]"Confirmatory data analysis is judicial or quasi-judicial in character." [4] Consider the following quote from page one of Tukey's text, "Exploratory data analysis is detective work—numerical detective work—or counting detective work—or graphical detective work. A detective investigating a crime needs both tools and understanding. ... Equally, the analyst of data needs both tools and understanding."

Another topic that must be considered carefully during exploratory causal analysis is hypothesis testing. Hypothesis (also called statistical or significance) testing is a standard part of modern Granger causality studies.[6] The early work of Granger was concerned with developing a "testable" definition of causality (see, e.g., [24, 25]), and by the late 1970s, hypothesis testing was a standard part of Granger causality studies (see, e.g., [26]). The same time period also saw many authors argue against the use of such testing in causality studies (see, e.g., [27, 28]). Statistical tests may or may not be useful for causal inference with time series causality tools, but the results of such tests should be not be interpreted beyond *exploratory* causal inference; i.e., the use of a hypothesis test does not imply a given causal inference is confirmatory rather than exploratory. For example, suppose a set of leanings[7] is created as part of the exploratory causal analysis. This set must be distilled somehow into a single causal inference. Suppose a null hypothesis N_0 is defined as "the mean of this set of leanings is negative" or $\langle \lambda_m \rangle < 0$. Let N_0 represent the causal inference $\mathbf{B} \rightarrow \mathbf{A}$ in a time series pair $(\mathbf{A}, \mathbf{B})$. If N_0 is rejected at significance level α, then the interpretation is that the causal inference $\mathbf{B} \rightarrow \mathbf{A}$ is rejected at significance level α, not that the causal inference $\mathbf{A} \rightarrow \mathbf{B}$ is "accepted" with confidence $1 - \alpha$. Furthermore, the test focuses on the rejection of a specific null hypothesis, and that null hypothesis is a statement about a specific calculation. The rejection of a given causal inference comes only from the causal interpretation of the calculation around which the null hypothesis was constructed. These points may seem obvious, but significance testing has led to confusion in the literature whereby the rejection of a given null hypothesis at a very low significance level is conflated with a high "confidence" in some causal interpretation [27, 28]. The statistical test is about a specific calculation on the data sets, and the analyst's confidence in a causal inference depends on concerns that usually are not addressed by such tests, including the assumptions used when performing the calculation (e.g., linear forecast models in a Granger causality calculation) and the validity and/or applicability of a given causal interpretation of that calculation. Statistical tests are also used in confirmatory causal analysis, for example by defining a null hypothesis of some model parameter being equal to a theoretically justified value. These tests are fundamentally different from those used in exploratory causal analysis because they rely on theories regarding the potential system dynamics. The point of such tests in confirmatory causal analysis is to *confirm* those theoretically justified models or parameter values. As emphasized previously, exploratory causal analysis does not *confirm* anything and does not rely on specific theories or models regarding the dynamics that generated the data being analyzed.

1.4   MOVING FORWARD WITH DATA CAUSALITY

The preface described three steps the field of causal studies needs to take for a practicing analyst to be able to apply the tools of the field in the manner of other statistical tools (such as, e.g., bootstrapping or confidence intervals). Those steps are

---

[6]See Section 3.1.
[7]See Section 3.5.

1. develop a loose taxonomy of the field to help frame the specific types of approaches an analyst may be seeking (e.g., time series causality)

2. carefully and deliberately divorce philosophy from data causality studies

3. present examples where the different approaches are compared on identical data sets, which have driving relationships that are strongly intuitive.

Step 1 has been attempted by several authors (see, e.g., [15, 16, 29]). However, the taxonomies are usually either too broad to help an analyst surveying the field for useful tools, or far too specific for such an analyst to understand without reading large amounts of background literature. This work introduces a taxonomy in Section 2 for which the category boundaries are defined by the type of question the analyst is asking and the type of data available. This approach necessarily leads to fuzzy boundaries within the taxonomy (e.g., many different types of data might be available to answer the same question), but the taxonomy is not meant to be rigorous. This type of taxonomy is currently missing from the field, and is intended to help guide the analyst within the vast literature of causality studies.

Step 2 is often ignored completely by most authors, with the notable exception of Granger [6]. Many authors often take the opposite approach and use philosophical arguments to justify a particular tool or set of tools (see, e.g., [15, 30, 31]). However, for a practicing analyst, this approach does not provide an understanding of how a particular tool might work with a given data set. The primary question for an analyst is "Does this tool work?" or, more specifically, "Does this particular approach to drawing causal inferences from the data do so correctly and consistently for the particular type of data available to me?" Many philosophical arguments require the analyst to define what they mean by "causal," "correctly," "consistently," and sometimes even "data," which can be both frustrating and potentially limiting for an analyst who may avoid a particular tool (that would otherwise be useful for them) because of philosophical objections. This work introduces the concept of exploratory causal analysis,[8] as opposed to confirmatory causal analysis, in which *operational* definitions of causality are presented for specific tools. Such causal definitions are specific and are not expected to meet any philosophical requirements such as those presented by Suppes [32] or Good [33].[9] Furthermore, this work posits that the data analysis may use many different operational definitions of causality to draw as much causal inference from the data as possible. This work not only outlines this new approach to data causality but also demonstrates its usefulness with different examples and illustrates how such an approach may be undertaken on a time series data set.[10]

Step 3 is, for an analyst looking for practical tools, a troubling omission of the current field. Time series causality studies in particular often introduce new tools using data sets for which there are not driving relationships that are either intuitively obvious or justified by outside theory

---

[8]See Section 1.3.
[9]See Section 1.2.
[10]See Section 4.

(see, e.g., [30, 34, 35]). It is difficult to interpret the results of such studies. Does the proposed tool work? Is the operational definition of causality presented in the study comparable to the causal inferences the analyst is hoping to draw from their own data? Such questions cannot be answered if the causal inferences implied by the proposed tools cannot be compared to nominally "correct" causal inferences (i.e., intuitive causal inferences or those supported by well-established theoretical models of the system dynamics). Many authors comment on how a proposed time series causality tool may compare to existing tools but rarely apply both tools to the same data, which leaves an analyst to rely on the authors' comments alone when trying to determine which tool might be best for a particular causal analysis. It has been shown that such an approach to introducing new time series causality tools can actually lead an analyst to trust that a given tool may provide reliable causal inference in situations where that is not true. Convergent cross-mapping (CCM)[11] was shown to provide counter-intuitive causal inferences for simple example systems, which this work addresses by introducing pairwise asymmetric inference; see Section 3.3 and [36]. It is shown in Section 4 that many different time series causality tools may provide identical causal inferences despite expectations to the contrary (e.g., compare the discussion of how Granger causality compares to CCM in [30] to the results presented in Section 4.2.5). This work introduces the idea of comparing different time series causality tools on identical data sets, not only to compare the tools but also to understand how applying different operational definitions of causality to the same data set may provide deeper insight into the data.

---

[11]See Section 3.3.

CHAPTER 2

Causality Studies

Studies of causality are as old as the scientific method itself.[1] An overview of such studies is far beyond the scope of this work and would probably fill several volumes. As Holland says, "So much has been written about causality by philosophers that it is impossible to give an adequate coverage of the ideas that they have expressed in a short article" [16]. Physicists, economists (and econometricians), mathematicians, statisticians, computer scientists, and scientists from a wide array of other fields (including medicine, psychology, and sociology) have also contributed greatly to the vast body of literature on causality.

This section is meant to provide a loose taxonomy of these studies, and is not meant to provide an overview of these studies. Authors often discuss causality with little attention given to the specific definitions of their causal language or the specific goals they are hoping to achieve with their work (see, e.g., [6, 16] for a discussion of such issues). This taxonomy is meant to explain where exploratory causal analysis, as it is defined in Section 1.3, relates to other causality studies. This taxonomy will not be complete, nor will the boundaries be clean. Several authors, particularly in theoretical physics, blur boundaries, e.g., between data studies and philosophy (see, e.g., [13]).

Holland outlines four primary focuses for authors studying causality, the ultimate meaningfulness of the notion of causality, the details of causal mechanisms, the causes of a given effect, and the effects of a given cause [16]. The first two goals in this list will fall into a category called *foundational causality*, and the last two will fall into a category called *data causality*. It is assumed that all causality studies can be placed into one of these two categories, but the categories are not considered disjoint. Data causality studies are characterized by the use of data (empirical or synthetic) in discussion of causality, and foundational causality studies are characterized as not being data causality studies.[2] Many authors have argued that foundational causality should not be studied independent of data causality (see, e.g., [7, 12, 16, 42–45]). The authors of data causality studies often bemoan the lack of academic consensus among foundational studies while recognizing the need for operational notions of causality in data analysis (see, e.g., [6, 15, 46–48]). Time

---

[1]Aristotle is often credited with both the first theory of causality [37, 38] and an early version of the scientific method [39] (along with other ancient Greek philosophers [40]). Aristotle's death was in 322 BCE [41].

[2]These two notions may seem at odds. If s is a causality study, then, given the description, it may belong to either the set of foundational studies, $\mathbb{F}$, or the set of data studies, $\mathbb{D}$; i.e., $s \in \mathbb{F}$ or $s \in \mathbb{D}$. If $\mathbb{F} \cap \mathbb{D} \neq \emptyset$, then $\exists s$ such that $s \in \mathbb{F}$ and $s \in \mathbb{D}$. If $\mathbb{F} = \mathbb{D}^c$, where $\mathbb{D}^c$ is the complement of $\mathbb{D}$, i.e., foundational studies are those studies that are not data studies, then $\exists s$ such that $s \in \mathbb{D}^c$ and $s \in \mathbb{D}$. This notion may seem nonsensical when presented in this set notation, but is common in practice due to the large variety of things that may be described as s. For example, a single study s may include the introduction of a completely new definition of "causality" and present several examples using empirical data to motivate the veracity of the definition. See the work of Illari et al. for an exploration of these ideas [5, 11].

series causality is considered a subset of data causality, and, therefore, data causality will be explored in more detail than foundational causality. Introductions and discussions of foundational causality studies can be found in [5, 9, 12, 13, 15, 31].

2.1   FOUNDATIONAL CAUSALITY

Do causes always precede effects? Is a cause required to precede an effect? How are causes and effects related in space-time? Is relativistic causality different from quantum causality and classical causality? What is quantum causality? Is causality a law of Nature? Is causality a principle used by practicing scientists? How are cause and effect perceived? Can there be differences in such perceptions without changing the underlying notion of causality? These questions are examples of foundational causality studies. Foundational causality, like all causality studies, is broadly inter-disciplinary. Three of the most prominent subfields are philosophical studies, natural science studies, and psychological studies. Legal causality (i.e., who is legally responsible for a given event) is often considered a part of the philosophical study of causality, albeit with a more practical, goal-oriented outlook (see, e.g., [49, 50]). A detailed introduction to legal causality can be found in [5].

2.1.1   PHILOSOPHICAL STUDIES

Aristotle developed a theory of causality that involved four types of causes: the material cause, the formal cause, the efficient cause, and the final cause [16, 37, 38, 47]. These ideas were refined by many philosophers, including Locke [51] and Decartes [52] in the 17th century and Kant [53] and Hume [54] in the 18th century. Hume's contribution to the philosophical discussion of causality is often considered among the greatest [5, 15, 16, 47], although disagreements with many of the fundamental postulates of Hume led to the relatively recent development of probabilistic causality by authors such as Suppes [32] and Good [33]. The basic notions of probabilistic causality have found favor among many data causality authors (see, e.g., [5, 31, 47]). Mill is often credited with developing the philosophical groundwork for drawing causal inferences from experiments [55] in the 19th century.

A reoccurring theme in philosophical studies of causality is the question of whether or not causality is (or should be) a part of reality (for some definition of "reality") [5, 47]. Many philosophers argue against causality as something that can be comprehended by human minds, including Plato in approximately 360 BCE [56], Spinoza in the 17th century [57], and Pearson in the 19th century [58]. Some philosophers have questioned the wisdom of making causality a requirement of reality, including Russel in the 20th century [20, 21]. The wide variety of different philosophical thoughts on causality has led some authors to question whether such questions can have answers at all (see, e.g., [11, 46, 59]).

2.1.2   NATURAL SCIENCE STUDIES

Causality, defined as cause must precede effect, is often considered a fundamental part of physics and other natural sciences [12, 13]. Many studies of causality in physics seek answers to questions regarding this so-called law of causality, including: can the law be violated?, can the law be derived from more fundamental notions of, e.g., information?, is the law universally applicable?, etc. These ideas are often framed in discussions of specific concepts and theories, including quantum interpretations (see, e.g., [60–62]), quantum irreversibility (see, e.g., [63]), quantum information (see, e.g., [64, 65]), astrophysics (see, e.g., [66, 67]), and relativity (see, e.g., [68–70]). Physicists in particular often study the relationship between causality and time (see, e.g., [71–74]).

Foundational causality studies are an active part of modern physics. Causal set theory, which relies on causality as a fundamental principle, has been proposed as a potentially fruitful approach to quantum gravity studies (see, e.g., [75, 76]). Closed timelike curves [77] have been used to make advances in quantum computational complexity [78] and may potentially limit the applicability of causality as a law of Nature (see, e.g., [79, 80]). There are even proposed quantum experiments to test "retrocausal" signaling [81]. However, some prominent physicists have argued, similar to Pearson's argument in statistics [58], that causal notions should not be a part of physics nor a concern of physicists [82]. Mach states, in his 1919 textbook *The Science of Mechanics*,

> "In speaking of cause and effect we arbitrarily give relief to those elements to whose connection we have to attend in the reproduction of a fact in the respect in which it is important to us. There is no cause and effect in nature; nature has but an individual existence; nature simply is. Recurrences of like cases in which *A* is always connected with *B*, that is, like results under like circumstances, that is again, the essence of the connection of cause and effect, exist but in the abstraction which we perform for the purpose of mentally reproducing the facts."[83]

Bunge provides a critical examination of Mach's proposal that causality is nothing more than functional dependence in [13].

2.1.3   PSYCHOLOGICAL (AND OTHER SOCIAL SCIENCE) STUDIES

In an overview of causality in psychological studies, White states, "psychology is concerned with how people understand and perceive causation, make causal inferences and attributions, and so forth" [84]. Psychologists are often concerned with how humans[3] reason causally (see, e.g., [87–89]). Psychologists have developed theories, e.g., power PC theory, "a causal power theory of the probabilistic contrast model" [90], to explain how a reasoner may become convinced that something causes something else.[4] Developmental psychology has sought to determine the minimum age at which a human may perceive causal relationships (see, e.g., [91–93]), and others have tried to determine if certain psychological illnesses may be linked to causal perceptions; e.g., are indi-

---

[3]Causal reasoning has also been studied in other species (see, e.g., [85, 86]).

[4]*Probabilistic contrast* (or *probability contrast*) is related to the causal penchant. See Section 3.5.

viduals who consistently perceive themselves to be the cause of some "bad" effect more likely to be depressed? (see, e.g., [94]). There has also been work to determine if perceptions of causality are related to cultural backgrounds (see, e.g., [95, 96]). A review of modern psychology studies of causal inference may be found in [97].

2.2   DATA CAUSALITY

Data causality focuses on determining the effects of given causes, or the causes of given effects, using data that represents measurements (of some kind) on a given system of interest. It is a fundamental part of science [1]. Rubin et al. have pointed out that most applications of statistics are concerned with questions of causality [9], and Winship et al. have argued that causality concerns are the motivation of much of the research that is conducted in the social sciences [10]. Data is primarily collected from one of two types of studies, experimental or observational [10]. Experimental studies are those studies in which the researcher directly intervenes into the system dynamics to change some aspect of the system, and observational studies are those studies in which such direct intervention by the researcher is not possible [9, 10, 15]. Many authors consider experimental studies to be the only way to truly determine causality in a system (see, e.g., [3, 16]). A researcher, however, may not be able to perform experiments on a given system for some reason, including ethical concerns (e.g., in medical studies) or technology limitations (e.g., in astrophysical studies). Thus, many data causality techniques were motivated by the researcher's potential restriction to only observational data [9, 10, 15].

Data causality tools encompass many different tools and techniques, including both experimental design (i.e., prior to data collection) and data analysis (i.e., post data collection). There is no clean taxonomy of these tools, so the following three categories are characterized primarily by the type of data being collected and/or analyzed. The first category, *statistical causality*, includes the most commonly used data causality methods, such as randomized trials. The second category, *computational causality*, includes computational approaches to large, usually observational, data sets. And, the third category, *time series causality*, includes techniques that rely on not only the collected data but also the time ordering of those data. Times series causality is, as mentioned in Section 1.2, the focus of this work.

Physics includes both experimental and observational data collection but is often more associated with foundational causality studies rather than data causality studies [12, 13, 74]. Data causality is, however, a large part of physics. Many of the time series causality tools discussed in this work were developed by physicists seeking solutions to questions of physical causality, including information-theoretic causality (see Section 3.2) and state space reconstruction causality (see Section 3.3). Physics is mentioned here to illustrate that despite the seemingly field-specific language used by many authors (e.g., "treatment" and "control"), data causality is a concern for many different scientific disciplines.

2.2.1   STATISTICAL CAUSALITY

Fisher outlined his views on what he called "the arts of experimental design" and "the valid interpretation of experimental results" in his influential book *The Design of Experiments* [3]. He introduced the notions of a null hypothesis, significance testing, and randomized trials. On the latter, he concludes, "It is apparent, therefore, that the random choice of the objects to be treated in different ways would be a complete guarantee of the validity of the test of significance, if these treatments were the last in time of the stages in the physical history of the objects which might affect their experimental reaction" [3]. Fisher's "randomized trials" design of an experiment is considered by many to be the only way to accurately draw causal inferences from experimental data [9, 10], although some authors disagree [15]. Fisher recognized that randomized trials would not always be possible in practice [3], and work was done by other authors to determine when (and how) causal inferences could be drawn from non-randomized trials. A detailed review of this work is in [10].

One of the most commonly used data causality approaches is the *counterfactual* or *potential outcome* approach. The label *counterfactual* has origins in philosophical studies of causality and is used to refer to outcomes of a cause that may presumably have occurred but did not [5, 9]. All such outcomes (both those that did and those that did not occur) are called potential outcomes, and unobserved potential outcomes are called counterfactual [10]. In the potential outcome framework, each unit u in the system may be exposed to one of two potential causes, the "treatment" t or the "control" c. The effect of that cause is given by some response variable Y acting on a given unit, i.e., $Y_t(u)$ is the effect of the treatment and $Y_c(u)$ is the effect of the control [9, 10, 16]. The "causal effect of t (relative to c) on u (as measured by Y)" [16] is

$$Y_t(u) - Y_c(u) \ . \tag{2.1}$$

The issue facing an analyst, however, is that Eq. (2.1) can never be measured for some systems. Holland refers to this as the *Fundamental Problem of Causal Inference*, i.e.,

Definition 2.1   *Fundamental Problem of Causal Inference*
It is impossible to *observe* the value of $Y_t(u)$ and $Y_c(u)$ on the same unit and, therefore, it is impossible to *observe* the effect of t on u. [16].

Holland posits two possible solutions to the fundamental problem of causal inference, the scientific solution and the statistical solution [16]. The scientific solution is the approach used in most experimental physics studies, i.e., if $Y_{(t,c)}(u_i, \tau_m)$ is the response variable Y (treatment or control) measured on unit u_i at time τ_m, then assume $Y_{(t,c)}(u_i, \tau_m) = Y_{(t,c)}(u_i, \tau_n) \ \forall \tau_m, \tau_n$ and/or assume $Y_{(t,c)}(u_i, \tau_m) = Y_{(t,c)}(u_j, \tau_m) \ \forall u_i, u_j$. Holland refers to the former as a temporal stability assumption (along with a "causal transience" assumption whereby it is assumed that the measurement at time τ_n does not affect measurements made at some time $\tau_m > \tau_n$) and the latter as a homogeneity assumption, but he points out that, in principle, neither assumption can be verified[5] [16]. The statistical solution is to define an *average causal effect*, T, as

$$T = E\left(Y_t - Y_c\right) = E\left(Y_t\right) - E\left(Y_c\right) \quad, \tag{2.2}$$

where $E(\cdot)$ is the expectation value [9, 10, 16]. Eq. (2.2) is often considered the great success of the potential outcome approach to data causality because, as Holland states, Eq. (2.2) "reveals that information on *different* units that *can be observed* can be used to gain knowledge about T" [16] (emphasis in original); i.e., causal inferences may be drawn from data despite the fundamental problem of causal inference.

Rubin is often credited with originally developing the potential outcome approach to data causality in [98] (see, e.g., [10, 16]), although Rubin himself, in [99], has claimed that the potential outcome framework was implicit in the early work of Fisher [3] and Neyman [100]. Many authors have contributed to the field since the work of Rubin; detailed overviews of the potential outcome model as an approach to data causality can be found in [9, 10], and an introduction to the philosophical framework of the approach can be found in [5].

The potential outcome approach, as outlined by Rubin and Holland [16, 98], has been criticized as a tool for causal inference, despite its apparently successful application to many problems in the social sciences [9, 10]. Dawid, in particular, has criticized the model as unnecessarily metaphysical, stating "… the counterfactual approach to causal inference is essentially metaphysical, and full of temptations to make 'inferences' that cannot be justified on the basis of empirical data and are thus unscientific" [101]. Dawid states directly, "There are many problems where workers who have grown familiar and comfortable with counterfactual modeling and analysis evidently consider that it forms the only satisfactory basis for *causal inference*. However, I have not as yet encountered any use of counterfactual models for inference about the effects of causes that is not either (a) a goat, delivering misleading inferences of no empirical content, or (b) interpretable, or readily reinterpretable, in noncounterfactual terms" [101] (emphasis in original).[6] Granger has also criticized the potential outcome approach for relying on an identification of "treatment" and "control" responses, which is often not possible in practice, and has labeled the approach as "cross-sectional causation" that cannot be applied to many important causal questions (particularly those regarding time series data) [103].

Dawid has introduced a decision-theoretic approach to data causality in contrast to the counterfactual approach [101, 104]. The decision-theoretic approach has been introduced using a similar framework, i.e., influence diagrams, to the directed acyclic graph (DAG) approach (which is discussed in the *computational causality* section) [104], but Dawid maintains that the decision-theoretic approach is distinct (and superior) to both the DAG and potential outcome (or counterfactual) approaches [105, 106]. Dawid extends the potential outcome approach by introducing

---

[5]"By careful work he may convince himself and others that this assumption is right, but he can never be absolutely certain" [16].

[6]Pearl, a proponent of the counterfactual approach, responds, "Dawid is correct in noting that many problems about the effects of causes can be reinterpreted and solved in noncounterfactual terms. Analogously, some of my colleagues can derive DeMoivre's theorem, $\cos ne = \mathrm{Re}[(\cos e + i \sin e)^n]$, without the use of those mistrustful imaginary numbers. So, should we strike complex analysis from our math books?" [102].

marginal distributions P_t and P_c to represent the uncertainty in the response Y (conditional on t or c) and a loss function $L(\cdot)$ to represent the consequence (or cost) of the decision to either apply the treatment t or the cost c to a unit u [101, 107].

Marked point processes have also been introduced as a method for drawing causal inferences from longitudinal studies [108, 109]. A set of causal events C_r with $r = 1, 2, \ldots, k$ is described by a pair of random variables (T_n, X_n) where T_n are the ordered time variables obeying $0 \leq T_1 \leq T_2 \leq \ldots$ and X_n are the "marks" that "describe" the causal events occurring at T_n [110]. The set of pairs $\{(T_n, X_n) \mid n \geq 1\}$ is known as a *marked point process* [110]. The marked points prior to some time t define a *history process* $H_t = \{(T_n, X_n) \mid T_n \leq t\}$, which may be used to define a *prediction process* to draw inferences regarding causal relationships within data sets. Marked point processes are used to draw causal inferences from data by modeling causal chains, which is similar to the directed acyclic graph approach of Pearl [15]. A concise introduction to causality studies using marked point processes can be found in [110].

2.2.2   COMPUTATIONAL CAUSALITY

Illari et al. remark in [11] that the 1980s saw a "revolution" in causality studies "stemming from interest amongst computer scientists and statisticians in probabilistic and graphical methods for reasoning with causal relationships." In 1997, identifying causal structure in data networks was identified as a major goal of machine learning [111]. Causal inference has, for some authors, become a computational data analysis problem (see, e.g., [15]). The research effort has become so popular in recent years that in 2011 a "virtual lab" was built for the express purpose of testing causal discovery algorithms [112].

Causal inference with structural equation modeling (SEM) relies on potential causal models fit to data and given causal interpretations [18, 113–115]. The method is similar to causal interpretation of regressions, which has been criticized by proponents of the potential outcome approach (see, e.g., [16]). Proponents of SEM, however, argue the methods are distinct from older regression techniques (see, e.g., [18]). Consider two data sets $\mathbf{M} = \{m_i\}$ and $\mathbf{N} = \{n_i\}$ where i is some general discrete index. An SEM of this system might be

$$m_i = \alpha + \beta_{mn} n_i + \varphi_i \ , \tag{2.3}$$

where β_{mn} is the structural coefficient (to which the causal interpretations are applied), α is a fitting coefficient, and φ_i are noise terms [18, 113, 115]. This equation is, by definition, an SEM, not a regression, which implies it is not equivalent [18] to

$$n_i = \beta_{mn}^{-1} (m_i - \alpha - \varphi_i) \ . \tag{2.4}$$

The first SEM is for the causal relationship $\mathbf{M} \to \mathbf{N}$, and modeling the associated causal relationship of this data pair, $\mathbf{N} \to \mathbf{M}$, requires a separate SEM in this framework, e.g.,

$$n_i = \gamma + \beta_{nm} m_i + \vartheta_i \ . \tag{2.5}$$

The ambiguity of the equals sign (i.e., "=") in SEM has been argued to be one of the primary sources of confusion for analysts using the method, and proposals have been made to instead use assignment operators (i.e., ":=") [114] or associated "path diagrams" (i.e., directed acyclic graphs) [15].

Bollen and Pearl provide an overview of many common criticisms (along with responses) of SEM [18]. The philosophical framework of using SEM for causal inference has also been criticized in the literature (see, e.g., [116, 117]). There have also been criticisms regarding the causal interpretations of SEM in particular (see, e.g., [118, 119]). It has been shown that SEM and the potential outcome approach are equivalent [10, 15, 18], although it has been argued that the potential outcome approach is more formal (see a summary of the argument in [18]), but that an SEM may be seen as a "formula for computing the potential outcomes" [120].

Graphical models are a part of many statistical fields, particularly statistical physics [121]. Path analysis has been a part of statistics since the early work of Wright [122]. Eerola notes in [110] that the practice actually began before Wright with the work of Yule in 1903 [123]. Directed acyclic graphs (DAGs; also called path diagrams or causal graphs) were originally an aid for causal inference with SEM [120]. The formal theory of DAGs was developed as part of machine learning efforts to model probabilistic reasoning, e.g., using Bayesian nets [124]. An example of a DAG for a system containing three variables A, B, and C might be

$$A \rightarrow B \rightarrow C \tag{2.6}$$

or

$$A \rightarrow B \leftarrow C \ , \tag{2.7}$$

where $\rightarrow$ and $\leftarrow$ represent general causal relations. Formally, a DAG obeys three conditions, the "causal Markov" condition, the "causal minimality" condition, and the faithfulness condition [15, 124]. It may be assumed that A and C are independent in the second DAG, which is written as $A \perp\!\!\!\perp C$. The Markov condition states A and C are also independent in the first DAG given (or "conditional on") the presence of B, i.e., $A \perp\!\!\!\perp C|B$ [124]. The minimality condition states that any proper subgraph (e.g., removing one of the directed edges, i.e., deleting an arrow) of the DAG containing the same vertex set will obey the Markov condition [124, 125]. The faithfulness condition states that all probabilistic independences within the model set of variables (e.g., $\{A, B, C\}$ in the example DAGs above) are required by the Markov condition [15, 124, 125]. These conditions are the framework by which a DAG models probabilistically causal relationships among variables [15].

SEM and DAG approaches to computational causality have been combined, along with counterfactuals, in the Structural Causal Model (SCM) proposed by Pearl [15, 126]. Pearl lists the goals of SCM as (in [126]):

1. "The unification of the graphical, potential outcome, structural equations, decision analytical [104], interventional [127], sufficient component [128] and probabilistic [32] approaches to causation; with each approach viewed as a restricted version of the SCM."

2. "The definition, axiomatization and algorithmization of counterfactuals and joint probabilities of counterfactuals."

3. "Reducing the evaluation of 'effects of causes,' 'mediated effects,' and 'causes of effects' to an algorithmic level of analysis."

4. "Solidifying the mathematical foundations of the potential-outcome model, and formulating the counterfactual foundations of structural equation models."

5. "Demystifying enigmatic notions such as 'confounding,' 'mediation,' 'ignorability,' 'comparability,' 'exchangeability (of populations),' 'superexogeneity' and others within a single and familiar conceptual framework."

6. "Weeding out myths and misconceptions from outdated traditions [129–134]."

Introductions to SCM may be found in [15, 126]. Pearl emphasizes that SCM pays special attention to confounding (through the "back-door criteria" for DAGs) and relies on non-parametric (i.e., in general, non-linear) SEMs to study counterfactuals, rather than the more traditional linear SEM [126]. Criticisms of SCM include arguments over the types of causes being explored with the approach and discussion of counterexamples for which SCM fails (see, e.g., [116, 135, 136]).

Kleinberg has proposed a logical approach to data causality whereby causes are formulated using probabilistic computation tree logic (PCTL) [31, 35, 137]. A potential cause c is identified in this framework as a PCTL formula obeying the following three conditions, c has some finite probability of occurring, the probability of the effect e is less than some value p, and there is a logical path from c to e that may occur with a probability greater than or equal to p within a finite amount of time (paraphrased from [137]). The causal significance ε_{avg} is defined as

$$\varepsilon_{avg}(c, e) = \sum_{x \in \mathcal{X} \setminus c} \frac{P(e|c \wedge x) - P(e|\neg c \wedge x)}{|\mathcal{X} \setminus c|} \quad , \tag{2.8}$$

where $\mathcal{X}$ is the set of all potential causes of e [35, 137], $P(a)$ is the probability of a, $A \wedge B$ is the intersection of A and B, and $A \setminus B$ is the set A minus the set B, i.e., $A \setminus B = \{x \mid x \in A \text{ and } x \notin B\}$. The logic used to define the cause and effect relationships is temporal, so there are time windows associated with each term of the sum in Eq. (2.8). A cause c is an ε-*significant cause* of e if $|\varepsilon_{avg}| \nless \varepsilon$ for some significance level ε [137]. A full introduction to this approach can be found in [31], which also includes discussions of algorithmically finding ε-significant causes in large data sets and appropriately determining ε for the given data (i.e., determining what significance level ε may lead to useful causal inference).

2.2.3   TIME SERIES CAUSALITY

Time series causality studies are data causality studies using time series data. Time series data are common to many research fields [138, 139], although many authors have argued that causal inferences should not be drawn from such data.[7] The rest of this work is dedicated to times series causality studies, specifically times series causality tools used in the exploratory causal analysis of bi-variate time series data.

---

[7]See Section 1.2 for a discussion of this point.

CHAPTER 3

Time Series Causality Tools

There are many time series causality tools. Attempting to catalog them all would be laborious, if it could be done at all, and this work would be out-of-date before it is ever printed. Most popular time series tools, however, appear to fall into one of five broad categories, Granger causality, information-theoretic causality, state space reconstruction causality, correlation causality, or penchant causality. These categories are defined by both the theoretical and computational differences between them, though there may be substantial theoretical, computational, and especially motivational similarities between them. Each category contains several different specific time series causality tools, and each category has critics and proponents, as described in the individual sections below. The main idea behind the exploratory causal analysis posited in this work is that tools from each of these categories are complementary to each other and can (and should) be used in concert during the analysis process.

The Granger causality tool category contains model-based tools. No other tool category relies on models of the time series data for causal inference. Granger's original motivation, as explained in Section 3.1, was independent any model considerations, including linearity and stationarity. However, Granger's insistence on a "operational definition" of causality [24] led to comparisons of potential forecast models for the data being investigated. The original operational definition proposed by Granger has been extended to include a wide variety of modeling frameworks, including linear, nonlinear, stationary, non-stationary, and computational methods,[1] available in many open-source software packages (see, e.g., [140–145]). In this work, any time series causality tool that involves models of the data is considered to belong to the Granger causality tool category. For example, both directed transfer function [146] and partial directed coherence [147] are considered tools that fall into this category.

The information-theoretic causality tool category contains tools that rely on information theory concepts for causal inferences;[2] i.e., all the tools in this category use entropies of some kind to quantify the information in the data [138]. Times series causality tools from this category are often lauded for the use of rigorous definitions of information and uncertainty (see, e.g., [148]) but are often computationally difficult and expensive to calculate on empirical data (see, e.g., [149]). There are many open source software packages available to calculate transfer entropy (see, e.g., [150–153]).

---

[1]See Section 3.1.
[2]See Section 3.2.

The state space reconstruction causality tool category contains tools that rely on state space reconstructions of the times series data [138, 154, 155] for causal inferences.[3] The state space reconstruction involves delay vector embedding that is often used in nonlinear and chaotic times series analysis (see, e.g., [156, 157]). The individual tools themselves, however, often differ significantly in the way embeddings are used for causal inferences (see Section 3.3). Embedding dimension and delay vector lag values must be set for each of the tools in this category, and this task can be difficult for empirical data (see, e.g., [158–161]). There is at least one open source software package for convergent cross mapping [162].

The correlation causality tool category contains tools that directly use correlation (or coherence) between the times series data for causal inference.[4] This category includes tools that rely on various signal processing techniques (e.g., whitening [139, 163] or Fourier transforms [139]) applied to the data before correlation calculations are used for causal inference, which may make this category appear to overlap with other categories. Even though convergent cross mapping involves a correlation calculation, the tool is considered part of the state space reconstruction category because the causal inference is posited to be impossible without the state space reconstruction. In practice, the tools in this category are all variants of lagged cross-correlations calculated in either the time or frequency domain.

The penchant causality tool category contains tools that draw causal inference from counting and comparing features between time series data sets.[5] Penchant causality and information-theoretic causality tools both must estimate probabilities from the data, which is not required for any of the other tool categories. Penchant causality tools, however, do not use information-theoretic concepts for causal inferences. The tools in this category rely only on identifying features in the time series pair that might be considered cause-effect pairs and then counting those pairs to estimate the associated probabilities. In this work, kernel density estimation [149] (and other methods for estimating probabilities from empirical data) will only be used with information-theoretic causality tools, not penchant causality tools. Instead, tolerance domains will be used with penchant causality tools to provide probability estimation methods unrelated to those used by the information-theoretic tools and to help ensure the exploratory causal analysis uses a time series causality tool set that is not unnecessarily homogeneous in calculation techniques. All the software used in this work is available online.[6]

3.1   GRANGER CAUSALITY

Clive Granger made the following comment in his 2003 Nobel Prize acceptance speech: "... causality has just two components:

1. The cause occurs before the effect; and

---

[3]See Section 3.3.
[4]See Section 3.4.
[5]See Section 3.5.
[6]https://github.com/jmmccracken

2. The cause contains information about the effect that is unique, and is in no other variable.

A *consequence* of these statements is that the causal variable can help forecast the effect variable after other data has first been used. . . ." (emphasis in original) [7]. This statement reflects the motivation behind Granger causality, which is perhaps the best known and most widely used time series causality tool. The precedent cause is common to all time series causality tools, in the way they are being considered in this work. Ganger does not provide formal definitions of "information," in contrast to the information-theoretic measures, but the second item in the quote may also be considered common to most time series causality tools. However, Granger's "consequence of these statements" is the formulation of a tool based on forecasting models, which is unique among the time series causality tools discussed in this work.

3.1.1   BACKGROUND

Consider a discrete universe with two time series $\mathbf{X} = \{X_t \mid t = 1, \ldots, n\}$ and $\mathbf{Y} = \{Y_t \mid t = 1, \ldots, n\}$, where $t = n$ is considered the present[7] time. All knowledge available in the universe at all times $t \leq n$ will be denoted as Ω_n. Granger introduces the following two axioms, which he assumes always hold [6] (Axioms A and B, *Ibid.*),

Axiom 3.1   The past and present may cause the future, but the future cannot cause the past.

Axiom 3.2 Ω_n contains no redundant information, so that if some variable $\mathbf{Z}$ is functionally related to one or more other variables, in a deterministic fashion, then $\mathbf{Z}$ should be excluded from Ω_n.

Given this framework,[8] Granger introduces the following definition: Given some set A, $\mathbf{Y}$ causes $\mathbf{X}$ if

$$P(X_{n+1} \in A|\Omega_n) \neq P(X_{n+1} \in A|\Omega_n - \mathbf{Y}) \ , \tag{3.1}$$

where $\Omega_n - \mathbf{Y}$ is defined as all the knowledge available in the universe *except* $\mathbf{Y}$; i.e., $\mathbf{Y}$ causes $\mathbf{X}$ if the probability that a future value of the series $\mathbf{X}$ is in some set A is different depending on whether or not all the knowledge in the universe up to the present is given or only that knowledge which is not in $\mathbf{Y}$ is given. Intuitively, this difference in probability implies knowledge of $\mathbf{Y}$ provides some knowledge of $\mathbf{X}$. Granger's original motivation was to take this definition and develop a practical, operational definition [6, 24, 25, 164].

The operational definition of Granger causality centers on building forecast models for the time series. For example, consider two autoregressive models for $f_1(\mathbf{X})$ and $f_2(\mathbf{X})$, one of which,

---

[7]The phrase "present" is used here to represent "now"; i.e., the time that separates what is considered to be the past from what is considered to be the future. These definitions are not meant to be formal in the physical sense (e.g., as a hyperplane in space-time). Rather, the language here is meant to be notional, as Granger originally used them [6].

[8]Granger also notes the following third axiom, *All causal relationships remain constant in direction throughout time* (Axiom C [6]), but this axiom is more relevant to drawing causal conclusions in the larger system from which $\mathbf{X}$ and $\mathbf{Y}$ have been measured. The goal of exploratory causal inference is, as described in Section 1.2, more limited in scope.

f_2, is multivariate; i.e.,

$$f_1(\mathbf{X}) = \left\{ X_t = \sum_{i=1}^{n} a_i X_{t-i} + \varepsilon_i \right\} \tag{3.2}$$

$$f_2(\mathbf{X}) = \left\{ X_t = \sum_{i=1}^{n} b_i X_{t-i} + \sum_{j=1}^{n} c_j Y_{t-j} + v_{ij} \right\}, \tag{3.3}$$

where ε, v are uncorrelated, Gaussian noise terms and a, b, and c are the model coefficients. If $\sigma^2(f_2(\mathbf{X})) < \sigma^2(f_1(\mathbf{X}))$, where $\sigma^2(\mathbf{Z})$ is the variance of the forecast errors for $\mathbf{Z}$, then $\mathbf{Y}$ "Granger causes" $\mathbf{X}$; i.e., if a better prediction of $\mathbf{X}$ can be made with $\mathbf{Y}$ rather than without it, then $\mathbf{Y}$ causes $\mathbf{X}$ in Granger's framework [27]. This operational definition corresponds Granger's original formulation of causality [24, 25], as Granger points out in [6]. Often, however, vector autoregressive models are used in practice, in conjunction with statistical tests (see, e.g., [165, 166]).

Consider a vector autoregressive (VAR) model for the system of $\mathbf{X}$ and $\mathbf{Y}$,

$$\begin{pmatrix} X_t \\ Y_t \end{pmatrix} = \sum_{i=1}^{n} \begin{pmatrix} A^i_{xx} & A^i_{xy} \\ A^i_{yx} & A^i_{yy} \end{pmatrix} \begin{pmatrix} X_{t-i} \\ Y_{t-i} \end{pmatrix} + \begin{pmatrix} \varepsilon_{x,t} \\ \varepsilon_{y,t} \end{pmatrix} \tag{3.4}$$

where $\varepsilon_{x,y}$ are, again, uncorrelated noise terms. Suppose the coefficient matrix A is found such that this model is optimal in some sense, e.g., the forecast error is minimized. If this model is the best fit possible for the system of $\mathbf{X}$ and $\mathbf{Y}$, then $\mathbf{Y}$ Granger causes $\mathbf{X}$ only if $A^i_{xy} \neq 0\ \forall i$ and $\mathbf{X}$ Granger causes $\mathbf{Y}$ only if $A^i_{yx} \neq 0\ \forall i$. A statistical test can then be formulated under the null hypothesis for non-causality, i.e., $\mathbf{Y}$ does not Granger cause $\mathbf{X}$ or vice versa, if $A^i_{xy} = 0$ or $A^i_{yx} = 0$. For example, Toda et al. outline the creation of several different statistics, including a Wald statistic, to test for Granger non-causality [167].

3.1.2   PRACTICAL USAGE

Consider a time series pair $\{\mathbf{X}, \mathbf{Y}\}$ with

$$\begin{aligned} \mathbf{X} &= \{X_t \mid t = 1, 2, \ldots, 10\} \\ &= \{0, 0, 1, 0, 0, 1, 0, 0, 1, 0\} \\ \mathbf{Y} &= \{Y_t \mid t = 1, 2, \ldots, 10\} \\ &= \{0, 0, 0, 1, 0, 0, 1, 0, 0, 1\}. \end{aligned}$$

The time series $\mathbf{Y}$ is described by $Y_t = X_{t-1}$ with the initial condition $Y_{t=1} = X_{t=1} = 0$. It follows that, intuitively, $\mathbf{X} \rightarrow \mathbf{Y}$. A model for this system can be written down, given the aforementioned initial condition, as

$$\begin{pmatrix} X_t \\ Y_t \end{pmatrix} = \begin{pmatrix} B_t & 0 \\ 1 & 0 \end{pmatrix} \begin{pmatrix} X_{t-1} \\ Y_{t-1} \end{pmatrix} + \begin{pmatrix} C_t \\ 0 \end{pmatrix}, \tag{3.5}$$

where $B_t = C_t = 0 \; \forall \, t \in \mathcal{T}$ and $B_t = C_t = 1 \; \forall \, t \notin \mathcal{T}$ with $\mathcal{T} = \{t \mid t \mod 3 \neq 0\}$. It follows that **X** Granger causes **Y** because $A_{21}^i \neq 0 \; \forall i$. The causal inference for this example is $\mathbf{X} \rightarrow \mathbf{Y}$, as expected.

In practice, fitting VAR models to the empirical data is not as straightforward via inspection as the example shown in Eq. (3.5). Often two different models,

$$\begin{pmatrix} X_t \\ Y_t \end{pmatrix} = \sum_{i=1}^{n} \begin{pmatrix} A_{xx,i} & A_{xy,i} \\ A_{yx,i} & A_{yy,i} \end{pmatrix} \begin{pmatrix} X_{t-i} \\ Y_{t-i} \end{pmatrix} + \begin{pmatrix} \varepsilon_{x,t} \\ \varepsilon_{y,t} \end{pmatrix} \tag{3.6}$$

(which is identical to Eq. (3.4)) and

$$\begin{pmatrix} X_t \\ Y_t \end{pmatrix} = \sum_{i=1}^{n} \begin{pmatrix} A'_{xx,i} & 0 \\ 0 & A'_{yy,i} \end{pmatrix} \begin{pmatrix} X_{t-i} \\ Y_{t-i} \end{pmatrix} + \begin{pmatrix} \varepsilon'_{x,t} \\ \varepsilon'_{y,t} \end{pmatrix} \; , \tag{3.7}$$

are fit to the data using well-known algorithms (see [143] for a discussion of such algorithms). These models can be compared using a null hypothesis that, e.g., $A_{xy,1} = A_{xy,2} = \ldots = A_{xy,n} = 0$ in Eq. (3.6), which can be tested using different test statistics [143, 167]. It has been argued that the log-likelihood statistic

$$F_{Y \rightarrow X} = \ln \frac{|\Sigma'_{xx}|}{|\Sigma_{xx}|} \; , \tag{3.8}$$

where $\Sigma'_{xx} = cov(\varepsilon'_{x,t})$ is the covariance of the **X** model residuals for Eq. (3.7) and $\Sigma_{xx} = cov(\varepsilon_{x,t})$ is the covariance of the **X** model residuals for Eq. (3.6), is the most appropriate Granger causality test statistic because of various formal mathematical properties and an information theoretic interpretation in analogy with transfer entropy [143, 168].

The interpretation of the magnitude of the log-likelihood statistic in information theoretic terms (i.e., as information flow) allows the individual statistics to be compared. As Barnett et al. point out [143], Granger causality statistics are often used only for hypothesis testing. However, the primary question of bivariate exploratory causal analysis is which time series may be seen as the stronger driver, and such a question may be difficult to address using traditional Granger causality testing. For example, consider a scenario in which the null hypothesis, N_{XY}, "**X** does not cause **Y**" has been rejected at a significance level of 0.05 and the null hypothesis, N_{YX}, "**Y** does not cause **X**" has also been rejected at a significance level of 0.05. What should the causal inference be? If N_{XY} is rejected at a lower significance level than N_{YX}, e.g., 10^{-3} and 0.05, respectively, should the causal inference be $\mathbf{Y} \rightarrow \mathbf{X}$ because there is "more confidence" that **X** does not cause **Y** than *vice versa* (i.e., N_{XY} is rejected at a lower significance level)? If the log-likelihood statistic is interpreted as an information flow, then the information flow between the time series can be compared.[9] For example, if $F_{X \rightarrow Y} - F_{Y \rightarrow X} > 0$, then the causal inference is $\mathbf{X} \rightarrow \mathbf{Y}$. See [143] for a discussion of interpreting the G ranger causality log-likelihood measure as a value with units (e.g., "bits"), as is done in this work, rather than just the test statistic of a hypothesis test.

---

[9]A similar approach will be used with the transfer entropy in Section 3.2.

Many of the theoretical criticisms of Granger causality can be levied on most of the other time series causality measures used in this work. For example, Granger emphasizes if **X** Granger causes **Y** then **X** should only be considered a potential physical cause [24, 169], which is true of every measure used in this work. The inability of Granger causality to identify confounding in the system, the apparent equating of causality with predictability, and the complete dependence on observed data for causal inference are, likewise, not unique to this tool. These issues are discussed in Section 1.2.

A pair of perceived limitations of Granger causality, linearity and stationarity [170], have been addressed by extensions of the operational definitions described in Section 3.1.1. Eq. (3.1) is not restricted by linearity or stationarity, but the operational definitions originally introduced by Granger relied on linear models for the times series under consideration. Nonlinear extensions have been approached with a variety of techniques, including SSR techniques [171], radial basis functions [172], and assumptions of local linearity [173]. The use of Granger causality with non-stationary time series has also been explored in the literature (see, e.g., [174–176]). Computational approaches to Granger causality have also been studied as tools for both causal modeling [177] and causal inference in general [178, 179]. The Granger causality framework has also been extended to include spectral (e.g., Fourier transform) methods [180].

3.2   INFORMATION-THEORETIC CAUSALITY

In 2000, Schreiber introduced transfer entropy as "an information theoretic measure . . . that quantifies the statistical coherence between systems evolving in time" [34]. The quantity has become one of the primary tools of information-theoretic times series causality [148, 149], which focuses on using the quantities used in information theory (i.e., entropy and mutual information) for causal inference between times series data.

3.2.1   BACKGROUND

Consider a random variable **X** that takes value X_n with probability p_n with $n = 1, 2, \ldots, N_X$. The probability distribution $P(\mathbf{X} = X_n) = p_n \, \forall X_n$ has a discrete Shannon entropy, H_X defined as

$$H_X = -\sum_{n=1}^{N_X} p_n \log_2 p_n \ ,$$

(3.9)

where the logarithm base determines the entropy units, which, in this case, is "bits" [34, 148, 149, 181], and $\log_2 0 := 0$. The Shannon entropy was developed as a measure of "of how uncertain we are of the outcome" $\mathbf{X} = X_n$ [181]; i.e., if a transmitter sends a message to a receiver over some channel, where the message is modeled as the random variable **X**, then how much certainty does the receiver have in receiving the specific value $\mathbf{X} = X_n$? If the transmitter only ever sends X_n, i.e., $P(\mathbf{X} = X_n) = 1$, then the receiver is completely certain that $\mathbf{X} = X_n$ and the Shannon entropy is zero [181].

The error made by incorrectly assuming $P(\mathbf{X} = X_n) = q_n$ (rather than p_n) is the Kullback entropy[10]

$$K_X = \sum_{n=1}^{N_X} p_n \log_2 \frac{p_n}{q_n} \;,$$ (3.10)

where, again, the base of the logarithm is due to the unit choice [34, 148, 149, 183]. Henceforth throughout this section, the logarithm base notation will be dropped and all logarithms should be assumed base 2 (i.e., everything is in units of bits) unless otherwise noted.

Consider a second random variable $\mathbf{Y}$ that takes value Y_m with probability p_m with $m = 1, 2, \ldots, N_Y$. The joint entropy is defined as

$$H_{X,Y} = -\sum_{n=1}^{N_X} \sum_{m=1}^{N_Y} p_{n,m} \log p_{n,m} \;,$$ (3.11)

where $p_{n,m}$ is the joint probability $P(\mathbf{X} = X_n, \mathbf{Y} = Y_m)$. If the two random variables are statistically independent, then $H_{X,Y} = H_X + H_Y$ [148], which is motivation for the introduction of the mutual information $I_{X;Y}$ as

$$I_{X;Y} = H_X + H_Y - H_{X,Y} = \sum_{n=1}^{N_X} \sum_{m=1}^{N_Y} p_{n,m} \log \frac{p_{n,m}}{p_n p_m} \;,$$ (3.12)

where the last equality can also be seen as the Kullback entropy due to assuming $P(\mathbf{X} = X_n, \mathbf{Y} = Y_m) = p_n p_m$ [149]. The mutual information is symmetric with respect to X and Y and is, therefore, not much use as a time series causality tool. Time lags and conditional entropies, i.e., $H_{X|Y} = H_{X,Y} - H_Y$, can be used to make the mutual information non-symmetric, but such modifications are not easily interpreted in terms of information flow [34] (the assumption is that a flow of information from one time series to another may be indicative of a driving relationship).

Schreiber's approach for using entropies to study dynamical structure of data was to focus on transitional probabilities in the system [34]. Suppose at time t, $\mathbf{X}(t) = X_n$ with some probability $P(\mathbf{X}(t) = X_n^{(t)}) = p_n$. The additional temporal structure allows for more complicated questions such as "what is the uncertainty in $\mathbf{X}(t)$ given some value of $\mathbf{X}(t-1)$?" The more basic question is how to define the transitional probabilities themselves, i.e., what is $P(\mathbf{X}(t) = X_n|\mathbf{X}(t-1) = X_{n-1}, \mathbf{X}(t-2) = X_{n-2}, \ldots, \mathbf{X}(0) = X_0) = p_{n|n-1,n-2,\ldots,0}$?

Assume $\mathbf{X}(t)$ is governed by a Markov process; i.e., $p_{n|n-1,n-2,\ldots,0} = p_{n|n-1}$ [184]. The basic concept of transfer entropy is to measure the error made by assuming the Markov process generating $\mathbf{X}$ does not depend at all on the second time series $\mathbf{Y}$. This error is quantified, in

---

[10]This quantity is also known as the Kullback-Leibler divergence, relative entropy, and discrimination information, among others. Kullback noted in 1987 that this quantity could be found under nine different names in the literature [182].

analogy to the mutual information expression shown above, with the Kullback entropy as

$$T_{Y \to X} = \sum_{n=1}^{N_X} \sum_{m=1}^{N_Y} p_{n+1,n,m} \log \frac{p_{n+1|n,m}}{p_{n+1|n}} \quad , \tag{3.13}$$

where $p_{n+1,n,m} = P(\mathbf{X}(t+1) = X_{n+1}, \mathbf{X}(t) = X_n, \mathbf{Y}(\tau) = Y_m)$ is the joint probability of $\mathbf{X}$ at times t and $t+1$ with $\mathbf{Y}$ at time τ, and the two conditional probabilities are $p_{n+1|n,m} = P(\mathbf{X}(t+1) = X_{n+1}|\mathbf{X}(t) = X_n, \mathbf{Y}(\tau) = Y_m)$ and $p_{n+1|n} = P(\mathbf{X}(t+1) = X_{n+1}|\mathbf{X}(t) = X_n)$. Schrieber notes that the "most natural choices" for τ are $\tau = t$ and $\tau = 0$ [34]. The quantity $T_{Y \to X}$ is called the transfer entropy. If $H_{X(t+1)|X(t)}$ is the conditional Shannon entropy at time $t+1$ given t, then the above expression can be rewritten as $T_{Y \to X} = H_{X(t+1)|X(t)} - H_{X(t+1)|X(t),Y(\tau)}$ [149]. Any pair of times series, $\mathbf{X}$ and $\mathbf{Y}$, will have (at least) two transfer entropies, $T_{Y \to X}$ and $T_{X \to Y}$, which can be compared to determine the times series causality [34].

Transfer entropy is often considered just one part of an information-theoretic approach to time series causality [148], but it will be the primary information-theoretic time series causality tool used in this work.

3.2.2   PRACTICAL USAGE

Consider a time series pair $\{\mathbf{X}, \mathbf{Y}\}$ with

$$\begin{aligned} \mathbf{X} &= \{X_t \mid t = 0, 1, \ldots, 9\} \\ &= \{0, 0, 1, 0, 0, 1, 0, 0, 1, 0\} \\ \mathbf{Y} &= \{Y_t \mid t = 0, 1, \ldots, 9\} \\ &= \{0, 0, 0, 1, 0, 0, 1, 0, 0, 1\} \, , \end{aligned}$$

where, for convenience, the notation is $Z_t := Z(t)$. This example system obeys the joint and conditional probabilities shown in Table 3.1. The causal intuition for this system is that $\mathbf{X}$ drives $\mathbf{Y}$.

The transfer entropy pair for this example can be calculated using Eq. (3.13) and Table 3.1 as

$$T_{Y \to X} = \frac{1}{9} \log\left(\frac{1}{2}\right) + \frac{2}{9} \log\left(2\right) + \frac{1}{3} \log\left(1\right) + \frac{1}{3} \log\left(\frac{3}{2}\right) \approx 0.31 \tag{3.14}$$

and

$$T_{X \to Y} = \frac{4}{9} \log\left(\frac{7}{4}\right) + \frac{2}{9} \log\left(1\right) + \frac{1}{3} \log\left(\frac{7}{3}\right) \approx 0.77 \; . \tag{3.15}$$

The difference $T_{X \to Y} - T_{Y \to X} = 0.46$ is positive, which implies $\mathbf{X} \to \mathbf{Y}$ as expected.

It can be shown that the transfer entropy is equivalent to Granger causality (up to a factor of 2) if $\mathbf{X}$ and $\mathbf{Y}$ are jointly multivariate Gaussian (i.e., both variables follow normal [Gaussian] distributions individually and jointly) [168]. As Barnett et al. point out in their paper, this result provided "for the first time a unified framework for data-driven causal inference that bridges

Table 3.1: Conditional and joint probabilities of $\{\mathbf{X}, \mathbf{Y}\}$ in Section 3.2.2 (the shorthand notation $Z_t := Z(t)$ is used for convenience)

$P(X_{t+1} = 0 \mid X_t = 0) = \frac{1}{2}$	$P(Y_{t+1} = 0 \mid Y_t = 0) = \frac{4}{7}$
$P(X_{t+1} = 1 \mid X_t = 0) = \frac{1}{2}$	$P(Y_{t+1} = 1 \mid Y_t = 0) = \frac{3}{7}$
$P(X_{t+1} = 0 \mid X_t = 1) = 1$	$P(Y_{t+1} = 0 \mid Y_t = 1) = 1$
$P(X_{t+1} = 1 \mid X_t = 1) = 0$	$P(Y_{t+1} = 1 \mid Y_t = 1) = 0$
$P(X_{t+1} = 0 \mid X_t = 0; Y_t = 0) = \frac{1}{4}$	$P(Y_{t+1} = 0 \mid Y_t = 0, X_t = 0) = 1$
$P(X_{t+1} = 0 \mid X_t = 0; Y_t = 1) = 1$	$P(Y_{t+1} = 0 \mid Y_t = 0, X_t = 1) = 0$
$P(X_{t+1} = 0 \mid X_t = 1; Y_t = 0) = 1$	$P(Y_{t+1} = 0 \mid Y_t = 1, X_t = 0) = 1$
$P(X_{t+1} = 0 \mid X_t = 1, Y_t = 1) = 0$	$P(Y_{t+1} = 0 \mid Y_t = 1, X_t = 1) = 0$
$P(X_{t+1} = 1 \mid X_t = 0, Y_t = 0) = \frac{3}{4}$	$P(Y_{t+1} = 1 \mid Y_t = 0, X_t = 0) = 0$
$P(X_{t+1} = 1 \mid X_t = 0, Y_t = 1) = 0$	$P(Y_{t+1} = 1 \mid Y_t = 0, X_t = 1) = 1$
$P(X_{t+1} = 1 \mid X_t = 1, Y_t = 0) = 0$	$P(Y_{t+1} = 1 \mid Y_t = 1, X_t = 0) = 0$
$P(X_{t+1} = 1 \mid X_t = 1, Y_t = 1) = 0$	$P(Y_{t+1} = 1 \mid Y_t = 1, X_t = 1) = 0$
$P(X_{t+1} = 0, X_t = 0, Y_t = 0) = \frac{1}{9}$	$P(Y_{t+1} = 0, Y_t = 0, X_t = 0) = \frac{4}{9}$
$P(X_{t+1} = 0, X_t = 0, Y_t = 1) = \frac{2}{9}$	$P(Y_{t+1} = 0, Y_t = 0, X_t = 1) = 0$
$P(X_{t+1} = 0, X_t = 1, Y_t = 0) = \frac{1}{3}$	$P(Y_{t+1} = 0, Y_t = 1, X_t = 0) = \frac{2}{9}$
$P(X_{t+1} = 0, X_t = 1, Y_t = 1) = 0$	$P(Y_{t+1} = 0, Y_t = 1, X_t = 1) = 0$
$P(X_{t+1} = 1, X_t = 0, Y_t = 0) = \frac{1}{3}$	$P(Y_{t+1} = 1, Y_t = 0, X_t = 0) = 0$
$P(X_{t+1} = 1, X_t = 0, Y_t = 1) = 0$	$P(Y_{t+1} = 1, Y_t = 0, X_t = 1) = \frac{1}{3}$
$P(X_{t+1} = 1, X_t = 1, Y_t = 0) = 0$	$P(Y_{t+1} = 1, Y_t = 1, X_t = 0) = 0$
$P(X_{t+1} = 1, X_t = 1, Y_t = 1) = 0$	$P(Y_{t+1} = 1, Y_t = 1, X_t = 1) = 0$

information-theoretic and autoregressive methods." The Gaussian restriction of the result implies the use of Granger causality and transfer entropy for exploratory causal analysis, in general, still depends on the data under consideration. Other authors have also explored the relationship between Granger causality and information-theoretic causality measures such as transfer entropy; e.g., see [185, 186].

Criticism of transfer entropy has centered on two major issues, the estimation of probabilities necessary for the calculation of transfer entropy in continuous processes [149, 187] and the inability of transfer entropy to quantify the strength of causal influences [188]. The latter is, as explained in Section 1.2, not a concern for time series causality as it has been defined in this work. The former is considered a more serious concern,[11] but transfer entropy has been shown to agree with intuitive notions of driving in several different examples, including both linear and nonlin-

---

[11]It is interesting to note that this concern seems to have first been raised by the author who originally introduced transfer entropy [149].

ear dynamics, using many different estimation techniques [149, 189]. Data is often assumed to be drawn from discrete state processes, in which case probability estimations are not a concern [149]. Estimating probabilities from the data is not unique to information-theoretic causality tools. These issues are also present in the leaning calculations described in Section 3.5. Possible issues involving probability estimations from the data will be addressed (following the philosophy outlined in Section 1.2) by not relying on transfer entropy alone to draw causal inferences and, if necessary, calculating the transfer entropy using different estimation techniques and comparing the results. Information flow has also been studied within a formal dynamical systems approach [190–192], which can lead to transfer entropy formulas that depend on fitting specific models to the data, rather than estimating the probabilities directly.

3.3   STATE SPACE RECONSTRUCTION CAUSALITY

State space reconstruction techniques have been used extensively in nonlinear (e.g., see [193]) and chaotic (e.g., see [156]) times series analysis. These techniques have a long history in physics [154, 156] and math [194]. The mathematical background of state space reconstruction, i.e., Takens' theorem, is explained in [155, 193]. State space reconstruction causality (SSRC) is closely related to simplex projection [195, 196], which predicts a point in the time series $\mathbf{X}$ at a time $t + 1$, labeled X_{t+1}, by using the points with the most similar histories to X_t. One of the most commonly used SSRC tools is convergent cross-mapping (CCM) [30], which was originally proposed for confirmatory, as opposed to exploratory, causal analysis (see, e.g., [30, 197]).[12] CCM will be the primary focus of this section. However, in practice, the primary SSRC tool used in the exploratory causal analysis discussed in this work will be pairwise asymmetric inference (PAI). It has been found that CCM may provide counter-intuitive causal inferences for simple dynamics with strong intuitive notions of causality; PAI was introduced to address this issue [36].

3.3.1   BACKGROUND

CCM uses points with the most similar histories to X_t to estimate Y_t, where similar histories are defined as nearest neighbors on a *shadow manifold*. The CCM correlation is the squared Pearson correlation coefficient[13] between the original time series $\mathbf{Y}$ and an estimate of $\mathbf{Y}$ made using its convergent cross-mapping with $\mathbf{X}$, which is labeled as $\mathbf{Y}|\tilde{\mathbf{X}}$:

$$C_{YX} = \left[\rho(\mathbf{Y}, \mathbf{Y}|\tilde{\mathbf{X}})\right]^2 \ . \tag{3.16}$$

Any pair of times series, $\mathbf{X}$ and $\mathbf{Y}$, will have two CCM correlations, C_{YX} and C_{XY}, which are compared to determine the time series causality. Sugihara et al. [30] define a difference of CCM

---

[12]For example, CCM has been used to draw conclusions regarding the "controversial sardine-anchovy-temperature" problem [30], confirm predictions of climate effects on sardines [198], compare the driving effects of precipitation, temperature, and solar radiation on the atmospheric CO_2 growth rate [199], and to quantify cognitive control in developmental psychology [200]. The technique has also been presented as a useful tool in studying the causality of respiratory systems in insects [201].
[13]This definition differs slightly from the definition in [30], which uses the un-squared Pearson's correlation coefficient.

correlations

$$\Delta = C_{YX} - C_{XY} \tag{3.17}$$

and use the sign of Δ (along with arguments of convergence[14]) to determine the time series causality between $\mathbf{X}$ and $\mathbf{Y}$. If $\mathbf{X}$ can be estimated using $\mathbf{Y}$ better than $\mathbf{Y}$ can be estimated using $\mathbf{X}$ (e.g., if $\Delta < 0$), then $\mathbf{X}$ drives $\mathbf{Y}$.

An algorithm to find the CCM correlations may be written in terms of five steps:

1. **Create the *shadow manifold* for X, called $\tilde{\mathbf{X}}$** Given an embedding dimension E, the *shadow manifold* of $\mathbf{X}$, labeled $\tilde{\mathbf{X}}$, is created by associating an E-dimensional vector (also called a *delay vector*) to each point X_t in $\mathbf{X}$, i.e., $\tilde{X}_t = (X_t, X_{t-\tau}, X_{t-2\tau}, \ldots, X_{t-(E-1)\tau})$. The first such vector is created at $t = 1 + (E-1)\tau$ and the last is at $t = L$ where L is the number of points in the time series (also called the *library length*).

2. **Find the nearest neighbors to a point in the shadow manifold at time t, $\tilde{X}_t$** The minimum number of points required for a bounding simplex in an E-dimensional space is $E + 1$ [195, 196]. Thus, the set of $E + 1$ nearest neighbors must be found for each point on the shadow manifold, $\tilde{X}_t$. For each $\tilde{X}_t$, the nearest neighbor search results in a set of distances that are ordered by closeness $\{d_1, d_2, \ldots, d_{E+1}\}$ and an associated set of times $\{\hat{t}_1, \hat{t}_2, \ldots, \hat{t}_{E+1}\}$. The distances from $\tilde{X}_t$ are

$$d_i = D\left(\tilde{X}_t, \tilde{X}_{\hat{t}_i}\right) \;, \tag{3.18}$$

where $D(\tilde{a}, \tilde{b})$ is the Euclidean distance between vectors $\tilde{a}$ and $\tilde{b}$.

3. **Create weights using the nearest neighbors** Each of the $E + 1$ nearest neighbors[15] are used to compute an associated weight. The weights are defined as

$$w_i = \frac{u_i}{N} \;, \tag{3.19}$$

where $u_i = e^{-d_i/d_1}$ and the normalization factor is $N = \sum_{j=1}^{E+1} u_j$.

4. **Estimate Y using the weights; (this estimate is called $\mathbf{Y}|\tilde{\mathbf{X}}$)** A point Y_t in $\mathbf{Y}$ is estimated using the weights calculated above. This estimate is

$$Y_t|\tilde{\mathbf{X}} = \sum_{i=1}^{E+1} w_i Y_{\hat{t}_i} \;. \tag{3.20}$$

5. **Compute the correlation between Y and $\mathbf{Y}|\tilde{\mathbf{X}}$** The CCM correlation is defined as

$$C_{YX} = \left[\rho\left(\mathbf{Y}, \mathbf{Y}|\tilde{\mathbf{X}}\right)\right]^2 \;, \tag{3.21}$$

where $\rho(A, B)$ is the standard Pearson's correlation coefficient between data sets A and B.

---

[14]Sugihara et al. consider convergence to be critically important for determining times series causality, and note that it is "a key property that distinguishes causation from simple correlation" [30].

[15]$E + 1$ is the minimum number of points needed for a bounding simplex in an E-dimensional space [195].

The CCM algorithm depends on the embedding dimension E and the lag time step τ. A dependence on E and τ is a feature of most state space reconstruction methods [158, 160, 202]. Sugihara et al. mention that "optimal embedding dimensions" are found using univariate SSR [30] (supplementary material), and other methods for determining E and τ for SSR algorithms can be found in the literature (e.g., [158, 160, 161]).

The CCM correlations shown above are not the only SSR correlations that can be tested. The time series $\mathbf{X}$ may also be estimated using a multivariate shadow manifold consisting of points from both $\mathbf{X}$ and $\mathbf{Y}$ [198]. For example, an $E + 1$ dimensional point in the a multivariate shadow manifold constructed using both $\mathbf{X}$ and $\mathbf{Y}$ may be defined as $\tilde{X}_t = (X_t, X_{t-\tau}, X_{t-2\tau}, \ldots, X_{t-(E-1)\tau}, Y_t)$. An estimate of $\mathbf{X}$ using weights from a shadow manifold using this specific construction will be referred to as $\mathbf{X}|(\mathbf{XY})$ and the correlation between this estimate and the original time series will be labeled $C_{X(XY)}$. A difference in CCM correlations similar to Δ can be defined using the multivariate embedding. Consider $\Delta' = C_{Y(YX)} - C_{X(XY)}$. This multivariate embedding extension of CCM is referred to as pairwise asymmetric inference (PAI). Comparisons between CCM and PAI can be found in [36].

3.3.2   PRACTICAL USAGE
Consider a time series pair $\{\mathbf{X}, \mathbf{Y}\}$ with

$$
\begin{aligned}
\mathbf{X} &= \{X_t \mid t = 1, 2, \ldots, 10\} \\
&= \{0, 0, 1, 0, 0, 1, 0, 0, 1, 0\} \\
\mathbf{Y} &= \{Y_t \mid t = 1, 2, \ldots, 10\} \\
&= \{0, 0, 0, 1, 0, 0, 1, 0, 0, 1\}.
\end{aligned}
$$

This noiseless impulse-response system obeys $Y_t = X_{t-1}$, leading to the intuitive causal relationship $\mathbf{X}$ drives $\mathbf{Y}$. The PAI shadow manifolds for these two signals can be written down using an embedding dimension $E = 3$ and a delay time step $\tau = 1$ as

$$
\tilde{\mathbf{X}} = \{\tilde{X}_t \mid t = 3, 4, \ldots, 10\} \tag{3.22}
$$

$$
= \left\{
\begin{array}{l}
\{1, 0, 0, 0\}, \\
\{0, 1, 0, 1\}, \\
\{0, 0, 1, 0\}, \\
\{1, 0, 0, 0\}, \\
\{0, 1, 0, 1\}, \\
\{0, 0, 1, 0\}, \\
\{1, 0, 0, 0\}, \\
\{0, 1, 0, 1\}
\end{array}
\right\} \tag{3.23}
$$

and

$$\tilde{\mathbf{Y}} = \{\tilde{Y}_t \mid t = 3, 4, \ldots, 10\} \tag{3.24}$$

$$= \left\{ \begin{array}{l} \{0, 0, 0, 1\}, \\ \{1, 0, 0, 0\}, \\ \{0, 1, 0, 0\}, \\ \{0, 0, 1, 1\}, \\ \{1, 0, 0, 0\}, \\ \{0, 1, 0, 0\}, \\ \{0, 0, 1, 1\}, \\ \{1, 0, 0, 0\} \end{array} \right\}. \tag{3.25}$$

Let the Euclidean distances between a PAI shadow manifold vector and every other vector in the manifold be

$$\{d_t^Z\} = \{D\left(\tilde{Z}_m, \tilde{Z}_t\right) \mid m \in [3, 10]\} \tag{3.26}$$

$$= \left\{\sqrt{(a_m - a_t)^2 + (b_m - b_t)^2 + (c_m - c_t)^2 + (d_m - d_t)^2} \mid m \in [3, 10]\right\} \tag{3.27}$$

where $\tilde{Z}_i = \{a_i, b_i, c_i, d_i\}$. The distances for the above shadow manifolds are

$$\{d_t^X\} = \left\{ \begin{array}{l} \{0, \sqrt{3}, \sqrt{2}, 0, \sqrt{3}, \sqrt{2}, 0, \sqrt{3}\}, \\ \{\sqrt{3}, 0, \sqrt{3}, \sqrt{3}, 0, \sqrt{3}, \sqrt{3}, 0\}, \\ \{\sqrt{2}, \sqrt{3}, 0, \sqrt{2}, \sqrt{3}, 0, \sqrt{2}, \sqrt{3}\}, \\ \{0, \sqrt{3}, \sqrt{2}, 0, \sqrt{3}, \sqrt{2}, 0, \sqrt{3}\}, \\ \{\sqrt{3}, 0, \sqrt{3}, \sqrt{3}, 0, \sqrt{3}, \sqrt{3}, 0\}, \\ \{\sqrt{2}, \sqrt{3}, 0, \sqrt{2}, \sqrt{3}, 0, \sqrt{2}, \sqrt{3}\}, \\ \{0, \sqrt{3}, \sqrt{2}, 0, \sqrt{3}, \sqrt{2}, 0, \sqrt{3}\}, \\ \{\sqrt{3}, 0, \sqrt{3}, \sqrt{3}, 0, \sqrt{3}, \sqrt{3}, 0\} \end{array} \right\} \tag{3.28}$$

and

$$\{d_t^Y\} = \left\{ \begin{array}{l} \{0, \sqrt{2}, \sqrt{2}, 1, \sqrt{2}, \sqrt{2}, 1, \sqrt{2}\}, \\ \{\sqrt{2}, 0, \sqrt{2}, \sqrt{3}, 0, \sqrt{2}, \sqrt{3}, 0\}, \\ \{\sqrt{2}, \sqrt{2}, 0, \sqrt{3}, \sqrt{2}, 0, \sqrt{3}, \sqrt{2}\}, \\ \{1, \sqrt{3}, \sqrt{3}, 0, \sqrt{3}, \sqrt{3}, 0, \sqrt{3}\}, \\ \{\sqrt{2}, 0, \sqrt{2}, \sqrt{3}, 0, \sqrt{2}, \sqrt{3}, 0\}, \\ \{\sqrt{2}, \sqrt{2}, 0, \sqrt{3}, \sqrt{2}, 0, \sqrt{3}, \sqrt{2}\}, \\ \{1, \sqrt{3}, \sqrt{3}, 0, \sqrt{3}, \sqrt{3}, 0, \sqrt{3}\}, \\ \{\sqrt{2}, 0, \sqrt{2}, \sqrt{3}, 0, \sqrt{2}, \sqrt{3}, 0\} \end{array} \right\}. \tag{3.29}$$

Let the pair $\left(d_{t(j)}^Z, j\right)$ denote for each manifold vector $\tilde{Z}_t$ the distance to $\tilde{Z}_j$ and the time $t = j$. The ordered distances, ignoring the self-distances, are then

$$\{d_t^X\}_o = \left\{ \begin{array}{l} \{(0,6),(0,9),\left(\sqrt{2},8\right),\left(\sqrt{2},5\right),\left(\sqrt{3},4\right),\left(\sqrt{3},7\right),\left(\sqrt{3},10\right)\}, \\ \{(0,7),(0,10),\left(\sqrt{3},3\right),\left(\sqrt{3},5\right),\left(\sqrt{3},6\right),\left(\sqrt{3},8\right),\left(\sqrt{3},9\right)\}, \\ \{(0,8),\left(\sqrt{2},3\right),\left(\sqrt{2},6\right),\left(\sqrt{2},9\right),\left(\sqrt{3},4\right),\left(\sqrt{3},7\right),\left(\sqrt{3},10\right)\}, \\ \{(0,3),(0,9),\left(\sqrt{2},5\right),\left(\sqrt{2},8\right),\left(\sqrt{3},4\right),\left(\sqrt{3},7\right),\left(\sqrt{3},10\right)\}, \\ \{(0,4),(0,10),\left(\sqrt{3},3\right),\left(\sqrt{3},5\right),\left(\sqrt{3},6\right),\left(\sqrt{3},8\right),\left(\sqrt{3},9\right)\}, \\ \{(0,5),\left(\sqrt{2},3\right),\left(\sqrt{2},6\right),\left(\sqrt{2},9\right),\left(\sqrt{3},4\right),\left(\sqrt{3},7\right),\left(\sqrt{3},10\right)\}, \\ \{(0,3),(0,6),\left(\sqrt{2},5\right),\left(\sqrt{2},8\right),\left(\sqrt{3},4\right),\left(\sqrt{3},7\right),\left(\sqrt{3},10\right)\}, \\ \{(0,4),(0,7),\left(\sqrt{3},3\right),\left(\sqrt{3},5\right),\left(\sqrt{3},6\right),\left(\sqrt{3},8\right),\left(\sqrt{3},9\right)\} \end{array} \right\} \tag{3.30}$$

and

$$\{d_t^Y\}_o = \left\{ \begin{array}{l} \{\left(\sqrt{2},4\right),\left(\sqrt{2},5\right),\left(\sqrt{2},7\right),\left(\sqrt{2},8\right),\left(\sqrt{2},10\right),(1,6),(1,9)\}, \\ \{(0,7),(0,10),\left(\sqrt{2},3\right),\left(\sqrt{2},5\right),\left(\sqrt{2},8\right),\left(\sqrt{3},6\right),\left(\sqrt{3},9\right)\}, \\ \{(0,8),\left(\sqrt{2},3\right),\left(\sqrt{2},4\right),\left(\sqrt{2},7\right),\left(\sqrt{2},10\right),\left(\sqrt{3},6\right),\left(\sqrt{3},9\right)\}, \\ \{(0,9),\left(\sqrt{3},4\right),\left(\sqrt{3},5\right),\left(\sqrt{3},7\right),\left(\sqrt{3},8\right),\left(\sqrt{3},10\right),(1,3)\}, \\ \{(0,4),(0,10),\left(\sqrt{2},3\right),\left(\sqrt{2},5\right),\left(\sqrt{2},8\right),\left(\sqrt{3},6\right),\left(\sqrt{3},9\right)\}, \\ \{(0,5),\left(\sqrt{2},3\right),\left(\sqrt{2},4\right),\left(\sqrt{2},7\right),\left(\sqrt{2},10\right),\left(\sqrt{3},6\right),\left(\sqrt{3},9\right)\}, \\ \{(0,6),\left(\sqrt{3},4\right),\left(\sqrt{3},5\right),\left(\sqrt{3},7\right),\left(\sqrt{3},8\right),\left(\sqrt{3},10\right),(1,3)\}, \\ \{(0,4),(0,7),\left(\sqrt{2},3\right),\left(\sqrt{2},5\right),\left(\sqrt{2},8\right),\left(\sqrt{3},6\right),\left(\sqrt{3},9\right)\} \end{array} \right\}. \tag{3.31}$$

These ordered distances are then subset to the $E + 1 = 4$ closest non-zero[16] distances for each vector in the manifold, which are then used to calculate the weights; i.e.,

$$
\{w_t^X\}_o = \left\{ \begin{array}{l}
\{(0.278, 8), (0.278, 5), (0.222, 4), (0.222, 7)\}, \\
\{(0.250, 3), (0.250, 5), (0.250, 6), (0.250, 8)\}, \\
\{(0.263, 3), (0.263, 6), (0.263, 9), (0.210, 4)\}, \\
\{(0.278, 5), (0.278, 8), (0.222, 4), (0.222, 7)\}, \\
\{(0.250, 3), (0.250, 5), (0.250, 6), (0.250, 8)\}, \\
\{(0.263, 3), (0.263, 6), (0.263, 9), (0.210, 4)\}, \\
\{(0.278, 5), (0.278, 8), (0.222, 4), (0.222, 7)\}, \\
\{(0.250, 3), (0.250, 5), (0.250, 6), (0.250, 8)\}
\end{array} \right\}
\tag{3.32}
$$

and

$$
\{w_t^Y\}_o = \left\{ \begin{array}{l}
\{(0.250, 4), (0.250, 5), (0.250, 7), (0.250, 8)\}, \\
\{(0.263, 3), (0.263, 5), (0.263, 8), (0.210, 6)\}, \\
\{(0.250, 3), (0.250, 4), (0.250, 7), (0.250, 10)\}, \\
\{(0.250, 4), (0.250, 5), (0.250, 7), (0.250, 8)\}, \\
\{(0.263, 3), (0.263, 5), (0.263, 8), (0.210, 6)\}, \\
\{(0.250, 3), (0.250, 4), (0.250, 7), (0.250, 10)\}, \\
\{(0.250, 4), (0.250, 5), (0.250, 7), (0.250, 8)\}, \\
\{(0.263, 3), (0.263, 5), (0.263, 8), (0.210, 6)\}
\end{array} \right\},
\tag{3.33}
$$

where $\{w_t^Z\}_o$ is the set of $E + 1$ pairs $\left(w_{t(j)}^Z, j \right)$ consisting of normalized[17] weights $w_{t(j)}$ calculated from the Euclidean distances between manifold vectors $\tilde{Z}_t$ and $\tilde{Z}_j$ as described in Eq. (3.19). These weights can be used to estimate the original signals as

$$
\begin{aligned}
\mathbf{Y}|\tilde{\mathbf{X}} &= \{Y_t|\tilde{\mathbf{X}} \mid t = 3, 4, \ldots, 10\} \tag{3.34} \\
&= \{0.444, 0, 0.210, 0.444, 0, 0.210, 0.444, 0\} \tag{3.35}
\end{aligned}
$$

and

$$
\begin{aligned}
\mathbf{X}|\tilde{\mathbf{Y}} &= \{X_t|\tilde{\mathbf{Y}} \mid t = 3, 4, \ldots, 10\} \tag{3.36} \\
&= \{0, 0.473, 0.250, 0, 0.473, 0.250, 0, 0.473\}. \tag{3.37}
\end{aligned}
$$

The PAI correlations are

$$
C_{YX} = \left[\rho\left(\mathbf{Y}, \mathbf{Y}|\tilde{\mathbf{X}} \right) \right]^2 \approx 0.78
\tag{3.38}
$$

---

[16]It is assumed that $d_1 \neq 0$ in Eq. (3.19) for any vector in the manifold because $\lim_{x \to \infty} e^{-x} = 0$ [184], which may imply that all the weights are zero if $d_1 = 0$. In practice (i.e., with empirical time series), d_1 is rarely identically equal to zero. However that is not always the case in simple examples like the one presented in this section. Only manifold vector pairs with non-zero distances are used in this example because it is only meant to be illustrative. The basic algorithm can be altered to handle zero distance manifold vector pairs (e.g., by changing how the weights are calculated), but such alterations would be counter to the point of this example.

[17]The weights are shown, for convenience, with only three significant digits. As such, not all of the normalized weights for a given manifold vector sum to one exactly.

and

$$C_{XY} = \left[\rho\left(\mathbf{X}, \mathbf{X}|\tilde{\mathbf{Y}}\right) \right]^2 \approx 0.88 \; , \tag{3.39}$$

which implies $\Delta \approx -0.1$ and $\mathbf{X} \to \mathbf{Y}$, as expected.

Proponents of SSRC often point out that such techniques may be more capable than Granger causality of identifying causality given weak or nonseparable dynamics [30, 203, 204]. SSRC techniques have been described as a "complementary approach" to the more standard time series causality studies using GC techniques [204]. One of the key ideas presented in this work is that no times series causality tool should be used alone, so we will follow the thought process implied by Cummins et al. in [204] and use both SSRC and GC techniques together.

It has been pointed out that CCM requires long times series [203], which may be a criticism of PAI as well.[18] Ma et al. have proposed a new SSRC technique called Cross Map Smoothness (CMS) to address this problem. CMS is a measure of the smoothness of the cross map (i.e., the map from, e.g., the shadow manifold of $\mathbf{Y}$ to the shadow manifold of $\mathbf{X}$). "Smoothness" is quantified by the training error of a neural network that takes, e.g., the shadow manifold of $\mathbf{Y}$ as input and the shadow manifold of $\mathbf{X}$ as output.[19] Clark et al. have also addressed the problem of using CCM with short time series by using regression techniques [206]. Other authors have argued that the projection back to a scalar time series from the state space reconstructions required by CCM and PAI (step 4 of the CCM algorithm) is "inconsistent" with the underlying theories of state space reconstruction [204]. Cummins et al. suggest "it is more consistent with the theory to test relationships between the state space reconstructions directly" and propose testing the continuity of the cross map[20] to determine the time series causality [204].

Fortunately, most of the time series used in this work are not short as Ma et al. defines it (i.e., 20 data points or less). This work is also not explicitly concerned with the consistency of a given technique with underlying SSR theories, especially if the technique is shown to be a reliable tool for causal inference. As such, the primary SSRC tool used in this work will be PAI, primarily because its reliability has been tested for the kinds of physical data sets explored in the following sections [36]. Of course, just as with any of the general classes of time series causality tools, other SSRC techniques can and should be used during the exploratory causal analysis if such techniques are thought to be useful for the given data.

---

[18]As Ma et al. state, "In fact, due to the computational way of state space reconstruction using delayed embedding technique, sufficiently long time series are required to guarantee that the nearest neighbors on the reconstructed attractor converge to the true neighborhood. ... Thus, detecting causality based on nearest neighbors and mutual neighbors essentially requires sufficiently long time series data to make reliable causality detection." PAI, like CCM, relies on finding nearest neighbors in the shadow manifold, so this concern also applies to PAI.

[19]This technique relies on the proof that any smooth map can be approximated by a neural network [205].

[20]This technique relies on a hypothesis-testing framework to explore continuity in state space reconstructions developed by Pecora et al. [207].

3.4   CORRELATION CAUSALITY

Lagged cross-correlation, also known as cross-lagged correlation, has been a popular time series causality tool in psychology [19, 208] and general signal analysis for many years [163]. Some authors consider it to be the first time series causality tool [208], with origins that can be traced back to 1901 [209]. The shortcomings of lagged cross-correlation have been discussed at length in the literature [19, 163, 210]. It is still, however, among the most popular time series causality tools because of its simplicity [19, 163].

3.4.1   BACKGROUND

Consider two time series $\mathbf{X} = \{X_t \mid t = 0, 1, 2, \dots, N\}$ and $\mathbf{Y} = \{Y_t \mid t = 0, 1, 2, \dots, N\}$. The lagged cross-correlation is defined as the normalized cross-covariance [139]

$$\rho_l^{xy} = \frac{E\left[(X_t - \mu_X)(Y_{t-l} - \mu_Y)\right]}{\sqrt{\sigma_X^2 \sigma_Y^2}} \;, \tag{3.40}$$

where l is the lag, σ_Z^2 is the variance of $\mathbf{Z}$, μ_Z is the mean of $\mathbf{Z}$, $E\left[(X_t - \mu_X)(Y_{t-l} - \mu_Y)\right]$ is the cross-covariance, and $E[z]$ is the expectation value of z [139, 163, 211]. Causal inference usually relies on using differences of these cross-correlations [19, 212, 213]; i.e.,

$$\Delta_l = |\rho_l^{xy}| - |\rho_l^{yx}| \;, \tag{3.41}$$

where $|z|$ is the absolute value of z. If Δ_l is positive, then the correlation between $\{X_t \mid t = l, l+1, l+2, \dots, N\}$ and $\{Y_t \mid t = 0, 1, 2, \dots, N - l\}$ is higher (i.e., further from zero) than the correlation between $\{X_t \mid t = 0, 1, 2, \dots, N - l\}$ and $\{Y_t \mid t = l, l+1, l+2, \dots, N\}$. The causal interpretation is as follows: If $\Delta_l > 0$, then $\mathbf{Y} \rightarrow \mathbf{X}$ at lag l, and if $\Delta_l < 0$, then $\mathbf{X} \rightarrow \mathbf{Y}$ at lag l. If $\Delta_l = 0$, then there is no causal inference at lag l. This interpretation depends on the definition of l as a lag, i.e., $l \leq 0$. If l is allowed to be a lead, i.e., $l > 0$, then these causal inference rules need to be altered.

3.4.2   PRACTICAL USAGE

Consider a time series pair $\{\mathbf{X}, \mathbf{Y}\}$ with

$$
\begin{aligned}
\mathbf{X} &= \{X_t \mid t = 0, 1, \dots, 9\} \\
&= \{0, 0, 1, 0, 0, 1, 0, 0, 1, 0\} \\
\mathbf{Y} &= \{Y_t \mid t = 0, 1, \dots, 9\} \\
&= \{0, 0, 0, 1, 0, 0, 1, 0, 0, 1\} \,.
\end{aligned}
$$

The structure of this system, i.e., $Y_t = X_{t-1}$, implies **X** drives **Y**. Table 3.2 can be calculated[21] by considering $l \in [0, 6]$.

Table 3.2: Lagged cross-correlation calculations for the example times series pair $\{\mathbf{X}, \mathbf{Y}\}$ in Section 3.4.2

l	ρ_l^{xy}	ρ_l^{yx}	Δ_l
0	0.43	0/43	0.0
1	0.38	1.0	-0.62
2	0.75	0.45	0.30
3	0.40	0.55	-0.15
4	0.32	1.0	-0.68
5	0.61	0.41	0.20
6	0.33	0.58	-0.24

One of the main practical issues with lagged cross-correlations is the decision of which lag to consider for causal inference. Table 3.2 shows Δ_l is nonzero for most l but does not suggest the same causal inference $\forall \ l > 0$. One strategy is to determine the relevant lag as the lag corresponding to the maximum difference, i.e., use Δ_m for causal inference where

$$|\Delta_m| = \max_l |\Delta_l| \ . \tag{3.42}$$

For this example, $\Delta_m = -0.68$, which implies $\mathbf{X} \to \mathbf{Y}$ as expected. However, $\Delta_m = \Delta_l$ at $l = 4$ not $l = 1$ as might be expected. One of the primary criticisms of lagged cross-correlation causality tools is the difficulties arising from auto-correlations in the data [163, 187, 210], as seen in this example. Other criticisms of lagged cross-correlations include underlying assumptions of linearity [163] and stationarity [19]. The simplicity of these calculations, however, suggest lagged cross-correlations should be calculated as a first step of an exploratory causal analysis.

3.5   PENCHANT CAUSALITY

Leaning is a time series causality technique that is derived directly from the definition of probabilistic causality [32]. Penchants, from which the leaning is found, are related to the Eells measure of causal strength (also called the probability contrast) [29], the weight of evidence [33], and the causal significance [31]. The main difference between penchants and the related measures is the intended use. Penchants and leanings will not be used as measures of causal strength in this

---

[21]Table 3.2 is calculated using the MATLAB function *corr*, which calculates the cross-correlation as $\rho_l^{xy} = m \sum_i \tilde{X}_i \tilde{Y}_{i-l}$ with $\tilde{Z}_t = Z_t - \mu_Z$ where $m = \left(||\tilde{\mathbf{X}}||_2 ||\tilde{\mathbf{Y}}||_2 \right)^{-1}$ is the inverse of the product of the Euclidean norms of $\tilde{\mathbf{X}}$ and $\tilde{\mathbf{Y}}$, μ_z is the mean of $\mathbf{Z}$, and $\tilde{\mathbf{Z}} = \{\tilde{Z}_t \mid t = 0, 1, \ldots, 9\}$.

work. Rather, they will be considered indicative of driving without any concern for notions of the "strength" of such driving. The penchant is non-parametric in the sense that there is no assumed functional relationship between the time series being investigated.

3.5.1 BACKGROUND

The causal penchant $\rho_{EC} \in [1, -1]$ is defined as

$$\rho_{EC} := P(E|C) - P(E|\bar{C}). \tag{3.43}$$

The motivation for this expression is in the straightforward interpretation of ρ_{EC} as a causal indicator; i.e., if C drives E, then $\rho_{EC} > 0$, and if $\rho_{EC} \leq 0$, then the direction of causal influence is undetermined.

One of the main ideas of the penchant definition is to circumvent philosophical issues regarding $P(E|\bar{C})$ as being unobservable (see [15]) by using an expression for Eq. (3.43) that does not have this term. Eq. (3.43) can be rewritten using Bayes' theorem

$$P(E|C) = P(C|E)\frac{P(E)}{P(C)} \tag{3.44}$$

and the definitions of probability complements

$$P(\bar{C}) = 1 - P(C) \tag{3.45}$$

$$P(\bar{C}|E) = 1 - P(C|E). \tag{3.46}$$

Using Eq. (3.46) with Eq. (3.44) gives

$$P(\bar{C}|E) = 1 - P(E|C)\frac{P(C)}{P(E)}$$

Inserting this into Eq. (3.44) written in terms of $\bar{C}$,

$$P(E|\bar{C}) = P(\bar{C}|E)\frac{P(E)}{P(\bar{C})}$$

yields an alternative form of the second term in Eq. (3.43)

$$P(E|\bar{C}) = \left(1 - P(E|C)\frac{P(C)}{P(E)}\right)\frac{P(E)}{1 - P(C)} ,$$

This expression gives a penchant that requires only a single conditional probability estimate:

$$\rho_{EC} = P(E|C)\left(1 + \frac{P(C)}{1 - P(C)}\right) - \frac{P(E)}{1 - P(C)} . \tag{3.47}$$

The penchant is not defined if $P(C)$ or $P(\bar{C})$ are zero (because the conditionals in Eq. (3.43) would be undefined). Thus, the penchant is not defined if $P(C) = 0$ or if $P(C) = 1$. The former condition corresponds to an inability to determine causal influence between two time series when a cause does not appear in one of the series; the latter condition is interpreted as an inability to determine causal influence between two time series if one is constant. The use of Bayes' theorem in the derivation of Eq. (3.47) implies that the penchant is not defined if $P(E)$ or $P(\bar{E})$ are zero. The method given in this work uses no *a priori* assignment of "cause" or "effect" to a given time series pair when using penchants for causal inference. So, operationally, the constraints on $P(C)$ and $P(E)$ only mean that the penchant is undefined between pairs of time series where one series is constant.

Consider the assignment of $\mathbf{X}$ as the cause, C, and $\mathbf{Y}$ as the effect, E. If $\rho_{EC} > 0$, then the probability that $\mathbf{X}$ drives $\mathbf{Y}$ is higher than the probability that it does not, i.e., $\mathbf{X} \xrightarrow{pen} \mathbf{Y}$. It is possible, however, that the penchant could also be positive when $\mathbf{X}$ is assumed as the effect and $\mathbf{Y}$ is assumed as the cause, i.e., $\mathbf{Y} \xrightarrow{pen} \mathbf{X}$. The leaning addresses this apparent confusion via

$$\lambda_{EC} := \rho_{EC} - \rho_{CE} \tag{3.48}$$

for which $\lambda_{EC} \in [-2, 2]$. A positive leaning implies the assumed cause C drives the assumed effect E more than the assumed effect drives the assumed cause, a negative leaning implies the effect E drives the assumed cause C more than the assumed cause drives the assumed effect, and a zero leaning yields no causal inference.

The possible outcomes are notated as

$$
\begin{aligned}
\lambda_{EC} > 0 \quad \{C, E\} = \{\mathbf{X}, \mathbf{Y}\} \quad &\Rightarrow \quad \mathbf{X} \xrightarrow{lean} \mathbf{Y} \\
\lambda_{EC} < 0 \quad \{C, E\} = \{\mathbf{X}, \mathbf{Y}\} \quad &\Rightarrow \quad \mathbf{Y} \xrightarrow{lean} \mathbf{X} \\
\lambda_{EC} = 0 \quad \{C, E\} = \{\mathbf{X}, \mathbf{Y}\} \quad &\Rightarrow \quad \text{no conclusion}
\end{aligned}
$$

with $\{C, E\} = \{\mathbf{A}, \mathbf{B}\}$ meaning $\mathbf{A}$ is the assumed cause and $\mathbf{B}$ as the assumed effect.

If $\lambda_{EC} > 0$ with $\mathbf{X}$ as the assumed cause and $\mathbf{Y}$ as the assumed effect, then $\mathbf{X}$ has a larger penchant to drive $\mathbf{Y}$ than $\mathbf{Y}$ does to drive $\mathbf{X}$. That is, $\lambda_{EC} > 0$ implies that the difference between the probability that $\mathbf{X}$ drives $\mathbf{Y}$ and the probability that it does not is higher than the difference between the probability that $\mathbf{Y}$ drives $\mathbf{X}$ and the probability that it does not.

The leaning is a function of four probabilities, $P(C)$, $P(E)$, $P(C|E)$, and $P(E|C)$. The usefulness of the leaning for causal inference will depend on an effective method for estimating these probabilities from times series and a more specific definition of the cause-effect assignment within the time series pair. An operational definition of C and E will need to be drawn directly from the time series data if the leaning is to be useful for causal inference. Such assignments, however, may be difficult to develop and may be considered arbitrary without some underlying theoretical support. For example, if the cause is x_{t-1} and the effect is y_t, then it may be considered unreasonable to provide a causal interpretation of the leaning without theoretical support that $\mathbf{X}$

may be expected to drive $\mathbf{Y}$ on the time scale of $\Delta t = 1$. This issue is, however, precisely one of the reasons for divorcing the causal inference proposed in this work (i.e., exploratory causal inference) from traditional ideas of causality. Statistical tools are associational, and cannot be given formal causal interpretation without the use of assumptions and outside theories (see [5] for an in-depth discussion of these ideas). In practice, many different potential cause-effect assignments may be used to calculate different leanings, which may then be compared as part of the causal analysis of the data. It can be noted that $\lambda_{AB} := \rho_{AB} - \rho_{BA} \Rightarrow -\lambda_{AB} = \rho_{BA} - \rho_{AB} := \lambda_{BA}$. Thus, the causal inference is independent of which times series is initially assumed to be the cause (or effect).

3.5.2   PRACTICAL USAGE

The initial derivation of the penchant and leaning is intentionally left in terms of the vague quantities of "cause" C and "effect" E. This decision comes directly from the probabilistic definition of causality as $P(E|C) > P(E|\bar{C})$ [32] (or, e.g., [5]), which was meant to convey a philosophical idea about the relationship between something capable of generally being called a cause, i.e., C, and something associated with it that is capable of generally being called an effect, i.e., E. Granger introduced his causality measure in a similar fashion using the same probabilistic causality definition [6]. The practicality of the penchant and leaning depends on an interpretation of how to make the terms C and E "operational" (in the sense that Granger defines the word [6]).

Consider a time series pair $(\mathbf{X}, \mathbf{Y})$ with

$$
\begin{aligned}
\mathbf{X} &= \{x_t \mid t = 0, 1, \ldots, 9\} \\
&= \{0, 0, 1, 0, 0, 1, 0, 0, 1, 0\} \\
\mathbf{Y} &= \{y_t \mid t = 0, 1, \ldots, 9\} \\
&= \{0, 0, 0, 1, 0, 0, 1, 0, 0, 1\} .
\end{aligned}
$$

Because $y_t = x_{t-1}$, one may conclude that $\mathbf{X}$ drives $\mathbf{Y}$. However, to show this result using a leaning calculation requires first a calculation using the cause-effect assignment $\{C, E\} = \{\mathbf{X}, \mathbf{Y}\}$. For consistency with the intuitive definition of causality, we require that a cause must precede an effect. It follows that a natural assignment may be $\{C, E\} = \{x_{t-l}, y_t\}$ for $1 \leq l < t \leq 9$. This cause-effect assignment will be referred to as the l-standard assignment.

The cause-effect assignment is an assignment of some countable feature of the data in one time series as the "cause" and another in the other time series as the "effect." For example, in the l-standard cause-effect assignment, the cause is the lag l time step in one time series and the effect is the current time step in the other. The leaning compares the symmetric application of these cause-effect definitions to the time series pair. So, for the above example of $\{C, E\} = \{x_{t-l}, y_t\}$, the first penchant will be calculated using $\{C, E\} = \{x_{t-l}, y_t\}$ and the second will be calculated using $\{C, E\} = \{y_{t-l}, x_t\}$. The second penchant is not the direct interchange of $C \leftrightarrow E$ from the first penchant because such an interchange would violate the assumption that a cause must precede an effect. For example, if the first penchant in the leaning calculation is calculated using $\{C, E\} = \{x_{t-l}, y_t\}$, then the second penchant is not calculated using $\{C, E\} = \{y_t, x_{t-l}\}$ because the definition of the effect, x_{t-l}, precedes the definition of the cause, y_t.

Given $(\mathbf{X}, \mathbf{Y})$, one possible penchant that can be defined using the 1-standard assignment is

$$\rho_{y_t=1, x_{t-1}=1} = \kappa \left(1 + \frac{P(x_{t-1} = 1)}{1 - P(x_{t-1} = 1)} \right) - \frac{P(y_t = 1)}{1 - P(x_{t-1} = 1)} ,$$

with $\kappa = P(y_t = 1 | x_{t-1} = 1)$. Another penchant defined using this assignment is $\rho_{y_t=0, x_{t-1}=0}$ with $\kappa = P(y_t = 0 | x_{t-1} = 0)$. These two penchants are called observed penchants because they correspond to conditions that were found in the measurements. Two other penchants have $\kappa = P(y_t = 0 | x_{t-1} = 1)$ and $\kappa = P(y_t = 1 | x_{t-1} = 0)$, and they are associated with unobserved conditions. These unobserved penchants in the leaning calculation would involve a comparison of how unlikely postulated causes are to cause given effects. Such comparisons are not as easily interpreted in the intuitive framework of causality, and as such, are often not used as part of leaning calculations.

The probabilities in the penchant calculations can be estimated from time series using counts, e.g.,

$$P(y_t = 1 | x_{t-1} = 1) = \frac{n_{EC}}{n_C} = \frac{3}{3} = 1 ,$$

where n_{EC} is the number of times $y_t = 1$ and $x_{t-1} = 1$ appears in $(\mathbf{X}, \mathbf{Y})$, and n_C is the number of times the assumed cause, $x_{t-1} = 1$, has appeared in $(\mathbf{X}, \mathbf{Y})$.

Estimating the other two probabilities in this penchant calculation using frequency counts from $(\mathbf{X}, \mathbf{Y})$ requires accounting for the assumption that the cause must precede the effect by shifting $\mathbf{X}$ and $\mathbf{Y}$ into $\tilde{\mathbf{X}}$ and $\tilde{\mathbf{Y}}$ such that, for any given t, $\tilde{x}_t$ precedes $\tilde{y}_t$. For this example, the shifted sequences are

$$\begin{aligned} \tilde{\mathbf{X}} &= \{0, 0, 1, 0, 0, 1, 0, 0, 1\} \\ \tilde{\mathbf{Y}} &= \{0, 0, 1, 0, 0, 1, 0, 0, 1\} \end{aligned}$$

which are both shorter than their counterparts above by a single value because the penchants are being calculated using the 1-standard cause-effect assignment. It follows that $\tilde{x}_t = x_{t-1}$ and $\tilde{y}_t = y_t$. The probabilities are then

$$P(y_t = 1) = \frac{n_E}{L} = \frac{3}{9} \tag{3.49}$$

and

$$P(x_{t-1} = 1) = \frac{n_C}{L} = \frac{3}{9} , \tag{3.50}$$

where n_C is the number of times $\tilde{x}_t = 1$, n_E is the number of times $\tilde{y}_t = 1$, and L is the ("library") length of $\tilde{\mathbf{X}}$ and $\tilde{\mathbf{Y}}$ (which are assumed to be the same length).

The two observed penchants in this example under the assumption that **X** causes **Y** (with $l = 1$) are

$$\rho_{y_t=1,x_{t-1}=1} = 1 \tag{3.51}$$

and

$$\rho_{y_t=0,x_{t-1}=0} = 1 \ .$$

The observed penchants when **Y** is assumed to cause **X** are

$$\rho_{x_t=1,y_{t-1}=0} = \frac{3}{7} \ ,$$
$$\rho_{x_t=0,y_{t-1}=1} = \frac{3}{7} \ ,$$

and

$$\rho_{x_t=0,y_{t-1}=0} = -\frac{3}{7} \ .$$

The mean observed penchant is the algebraic mean of the observed penchants. For **X** causes **Y**, it is

$$\begin{aligned}
\langle \rho_{y_t,x_{t-1}} \rangle &= \frac{1}{2} \left(\rho_{y_t=1,x_{t-1}=1} + \rho_{y_t=0,x_{t-1}=0} \right) \\
&= 1
\end{aligned}$$

and for **Y** causes **X** is

$$\begin{aligned}
\langle \rho_{x_t,y_{t-1}} \rangle &= \frac{1}{3} \big(\rho_{x_t=1,y_{t-1}=0} \\
&\quad + \rho_{x_t=0,y_{t-1}=1} + \rho_{x_t=0,y_{t-1}=0} \big) \\
&= \frac{1}{7} \ .
\end{aligned}$$

The mean observed leaning that follows from the definition of the mean observed penchants is

$$\begin{aligned}
\langle \lambda_{y_t,x_{t-1}} \rangle &= \langle \rho_{y_t,x_{t-1}} \rangle - \langle \rho_{x_t,y_{t-1}} \rangle \tag{3.52} \\
&= \frac{6}{7} \ . \tag{3.53}
\end{aligned}$$

The weighted mean observed penchant is defined similarly to the mean observed penchant, but each penchant is weighted by the number of times it appears in the data; e.g.,

$$\begin{aligned}
\langle \rho_{y_t,x_{t-1}} \rangle_w &= \frac{1}{L} \big(n_{y_t=1,x_{t-1}=1} \rho_{y_t=1,x_{t-1}=1} \\
&\quad + n_{y_t=0,x_{t-1}=0} \rho_{y_t=0,x_{t-1}=0} \big) \\
&= 1
\end{aligned}$$

and

$$
\begin{aligned}
\langle \rho_{x_t, y_{t-1}} \rangle_w &= \frac{1}{L} \big(n_{x_t=1, y_{t-1}=0} \rho_{x_t=1, y_{t-1}=0} \\
&\quad + n_{x_t=0, y_{t-1}=1} \rho_{x_t=0, y_{t-1}=1} \\
&\quad + n_{x_t=0, y_{t-1}=0} \rho_{x_t=0, y_{t-1}=0} \big) \\
&= \frac{3}{63} ,
\end{aligned}
$$

where $n_{a,b}$ is the number of times the assumed cause a appears with the assumed effect b and L is the library length of $\tilde{\mathbf{X}}$ (i.e., $L = N - l$ where N is the library length of $\mathbf{X}$ and l is the lag used in the l-standard cause-effect assignment).

The weighted mean observed leaning follows naturally as

$$
\begin{aligned}
\langle \lambda_{y_t, x_{t-1}} \rangle_w &= \langle \rho_{y_t, x_{t-1}} \rangle_w - \langle \rho_{x_t, y_{t-1}} \rangle_w \\
&= \frac{60}{63} .
\end{aligned}
$$

For this example, $\langle \lambda_{y_t, x_{t-1}} \rangle_w \Rightarrow \mathbf{X} \to \mathbf{Y}$ as expected.

Conceptually, the weighted mean observed penchant is preferred to the mean penchant because it accounts for the frequency of observed cause-effect pairs within the data, which is assumed to be a predictor of causal influence. For example, given some pair $(\mathbf{A}, \mathbf{B})$, if it is known that a_{t-1} causes b_t and both $b_t = 0 \mid a_{t-1} = 0$ and $b_t = 0 \mid a_{t-1} = 1$ are observed, then comparison of the frequencies of occurrence is used to determine which of the two pairs represents the cause-effect relationship.

If the example time series contained noise, then a realization of the example time series $(\mathbf{X}', \mathbf{Y}')$ could be

$$
\begin{aligned}
\mathbf{X}' &= \{ x_t' \mid t = 0, 1, \ldots, 9 \} \\
&= \{ 0, 0, 1.1, 0, 0, 1, -0.1, 0, 0.9, 0 \} \\
\mathbf{Y}' &= \{ y_t' \mid t = 0, 1, \ldots, 9 \} \\
&= \{ 0, -0.2, 0.1, 1.2, 0, 0.1, 0.9, -0.1, 0, 1 \} .
\end{aligned}
$$

The previous time series pair, $(\mathbf{X}, \mathbf{Y})$, had only five observed penchants, but $(\mathbf{X}', \mathbf{Y}')$ has more due to the noise. It can be seen in the time series definitions that $x_t' = x_t \pm 0.1 := x_t \pm \delta_x$ and $x_t' = x_t \pm 0.2 := x_t \pm \delta_y$. The weighted mean observed leaning for $(\mathbf{X}', \mathbf{Y}')$ is $\langle \lambda_{y_t', x_{t-1}'} \rangle_w \approx 0.19$.

If the noise is not restricted to a small set of discrete values, then the effects of noise on the leaning calculations can be addressed by using the tolerances δ_x and δ_y in the probability estimations from the data. For example, the penchant calculation in Eq. (3.51) relied on estimating $P(y_t = 1 \mid x_{t-1} = 1)$ from the data, but if, instead, the data is known to be noisy, then the relevant probability estimate may be $P(y_t \in [1 - \delta_y, 1 + \delta_y] \mid x_{t-1} \in [1 - \delta_x, 1 + \delta_x])$. If the tolerances, δ_x and δ_y, are made large enough, then the noisy system weighted mean observed leaning, $\langle \lambda_{y_t' \pm \delta_y, x_{t-1}' \pm \delta_x} \rangle_w$, can, at least in the simple examples considered here, be made equal to the noiseless system weighted mean observed leaning, i.e., $\langle \lambda_{y_t' \pm \delta_y, x_{t-1}' \pm \delta_x} \rangle_w = \langle \lambda_{y_t, x_{t-1}} \rangle_w$. Tolerance domains, however, can be set too large. If the tolerance domain is large enough to encompass every point in the time series, then the probability of the assumed cause becomes one, which leads to undefined penchants. For example, given the symmetric definition of the tolerance domain used here, $\delta_x = 2$ implies $P(x_{t-1} = 1 \pm \delta_x) = 1$, which implies $\langle \lambda_{y_t', x_{t-1}} \rangle_w$ is undefined. The tolerance domains can be interpreted as the set of values that an analyst is willing to consider equivalent causes or effects. For example, if the lag l time step of $\mathbf{X}$, i.e., x_{t-l}, is the assumed cause of some assumed effect (e.g., the current time step of $\mathbf{Y}$, y_t), then an x-tolerance domain of $[x_{t-1} - a, x_{t-1} + b]$ may be thought of as an analyst's willingness to consider all values that fall within that domain equivalently as the assumed cause of that assumed effect.

Reasonable leaning calculations require an understanding of the noise in the measurements, which may not always be possible. Estimating relevant tolerance domains is one of the two key difficulties in using penchants and leanings for causal inference. The other is finding an appropriate cause-effect assignment.

CHAPTER 4

Exploratory Causal Analysis

Section 1.3 outlined a proposed framework for exploratory causal analysis. This framework was characterized more by an approach to the analysis (i.e., a way of thinking about the data analysis) rather than the use of any specific technique. This work will focus on performing exploratory causal analysis using a report of five different time series causality tools, one from each category presented in Section 3.

Leanings will be used with various cause-effect assignments and tolerance domains, which will be set based on the data being analyzed. The weighted mean observed leanings will be calculated using MATLAB.[1] Lagged cross-correlation differences will be used with various lags but without any pre-treatment of the data. All lagged cross-correlations will be calculated using the standard MATLAB function *corr*. PAI will be calculated using the methods and software[2] discussed in Section 3.3; the embedding dimension will always be $E = 3$ and the delay time step will always be $\tau = 1$, unless otherwise noted. Transfer entropy will be calculated using the Java Information Dynamics Toolkit (JIDT) [153]. This software package includes different methods for estimating probabilities from the data, but, unless otherwise noted, this work will only calculate the transfer entropy using the default kernel estimator with normalized data, a delay time step of 1, and a kernel width of 0.5.[3] Granger causality will be calculated using the Multivariate Granger Causality (MVGC) toolbox, which calculates Granger causality in both the temporal and spectral domains [176]. The MVGC toolbox provides several different ways to estimate the model order for the VAR model estimation step of the Granger causality calculation. For consistency, the VAR model order will be set using the Bayesian information criterion [139, 214] routine provided by the MVGC toolbox [176], unless otherwise noted.

4.1   ECA SUMMARY

Exploratory causal analysis is the first step in a more rigorous data causality study of a system. The results should guide future experiments or more formal and rigorous data techniques. Nevertheless, there is a use in having a quick causal summary for a given time series pair. Consider a

---

[1]The scripts and functions used to calculate the penchants and leanings can be found at https://github.com/jmmccracken.
[2]The software used to calculate PAI can be found at https://github.com/jmmccracken.
[3]I.e., unless otherwise noted, the JIDT transfer entropy calculator will always be instantiated in MATLAB with the following commands,
“`teCalc=javaObject('infodynamics.measures.continuous.kernel.TransferEntropyCalculatorKernel');`,”
“`teCalc.setProperty('NORMALISE','true');`,” and `teCalc.initialize(1,0.5);`.”

ternary vector defined as

$$\vec{g} = (g_1, g_2, g_3, \ldots, g_n) \ , \tag{4.1}$$

where each trit[4] represents the causal inference of a given time series causality tool for a given time series pair $(\mathbf{X}, \mathbf{Y})$, 0 for $\mathbf{X} \to \mathbf{Y}$, 1 for $\mathbf{Y} \to \mathbf{X}$, and 2 if there is no conclusion (e.g., if the leaning is 0 within some expected error for all the tested cause-effect assignments and tolerance domains). The vector $\vec{g}$ may then be used to provide a concise summary of the exploratory causal analysis (ECA) results, which will be called the *ECA summary*; i.e., if all $g_i = 0$, then $\mathbf{X} \to \mathbf{Y}$ and if all $g_i = 1$, then $\mathbf{Y} \to \mathbf{X}$. The inner product of $\vec{g}$ with itself may be a simple test for an ECA summary,

$$\vec{g} \cdot \vec{g} = |\vec{g}|^2 = \begin{cases} 0 & \text{ECA summary is } \mathbf{X} \to \mathbf{Y} \\ \text{anything else} & \text{ECA summary is not defined or } \mathbf{Y} \to \mathbf{X} \end{cases} \tag{4.2}$$

If $g_i = 1 \ \forall g_i \in \vec{g}$, then the ECA summary would be $\mathbf{Y} \to \mathbf{X}$, which may make it tempting to interpret $|\vec{g}|^2 = n$ as implying $\mathbf{Y} \to \mathbf{X}$. It may be true, however, that $|\vec{g}|^2 = n$ but $g_i \neq 1 \ \forall g_i \in \vec{g}$. As discussed in the previous section, $n = 5$ in all the examples in this work.

It should be emphasized that the ECA summary is neither the final product nor the only conclusion that should be drawn from the exploratory causal analysis. The ECA summary will often be undefined in situations where a majority of g-trits in $\vec{g}$ agree with each other. For example, the transfer entropy, Granger causality, lagged cross-correlation, and PAI tools could all provide a causal inference of $\mathbf{Y} \to \mathbf{X}$ while the leaning fails to provide any causal inference because, e.g., the cause-effect assignment is inappropriate for the system being studied. An undefined ECA summary does not immediately imply that the exploratory causal inference is inconclusive. Rather, it only implies that each time series causality tool may need to be applied more carefully.

The automated generation of ECA summaries for a set of time series may be appealing in the speed and ease with which a large number of causal inferences can be performed, but such automated procedures should not be considered a substitution for a more complete exploratory causal analysis. The ECA summary is part of exploratory causal analysis and, as discussed in Section 1.3, no part of such analysis should be confused with causality as it is defined traditionally in fields such as physics and philosophy.

In all the examples presented in this work, $\vec{g}$ has five elements with g_1 as the causal inference implied by the JIDT transfer entropy, g_2 as the causal inference implied by the MVGC Granger causality log-likelihood test statistics, g_3 as the causal inference implied by the PAI differences, g_4 as the causal inference implied by the weighted mean observed leanings averaged over all the tested lags, and g_5 as the causal inference implied by the lagged cross-correlation differences averaged over all the tested lags.

---

[4]*trinary dig*it (see, e.g., [215]).

4.2   SYNTHETIC DATA EXAMPLES

Synthetic data sets can be used to explore how the results of an exploratory causal analysis compare with the intuitive causal structure of the system. Simple systems with strongly intuitive causal structure are useful in demonstrating the strengths and weaknesses of different approaches to the analysis. Example systems with less intuitive causal structure will also be used because of their relationship to physical systems, which allows outside theories to be used as confirmation of the exploratory causal analysis results.

4.2.1   IMPULSE WITH LINEAR RESPONSE

Consider the linear example dynamical system of

$$\{\mathbf{X}, \mathbf{Y}\} = \{\{x_t\}, \{y_t\}\} \tag{4.3}$$

where $t = 0, 1, \ldots, L$,

$$x_t = \begin{cases} 2 & t = 1 \\ A\eta_t & \forall\, t \in \{t \mid t \neq 1 \text{ and } t \bmod 5 \neq 0\} \\ 2 & \forall\, t \in \{t \mid t \bmod 5 = 0\} \end{cases}$$

and

$$y_t = x_{t-1} + B\eta_t$$

with $y_0 = 0$, $A, B \in \mathbb{R} \geq 0$ and $\eta_t \sim \mathcal{N}(0, 1)$. Specifically, consider $A \in [0, 1]$ and $B \in [0, 1]$. The driving system $\mathbf{X}$ is a periodic impulse with a signal amplitude above the maximum noise level of both the driving and the response systems, and the response system $\mathbf{Y}$ is a lagged version of the driving signal with standard normal (i.e., $\mathcal{N}(0, 1)$) noise of amplitude B applied at each time step.

Let the instance of Eq. (4.3) with $L = 500$, $A = 0.1$, and $B = 0.4$ shown in Figure 4.1 be considered the synthetic data set $(\mathbf{X}, \mathbf{Y})$ upon which the exploratory causal analysis will be performed. A preliminary visual inspection of Figure 4.1 shows $\mathbf{X}$ appears less noisy than $\mathbf{Y}$ (as expected), but perhaps more importantly, both times series appear to have two major groupings for their data, around 0 and 2 for $\mathbf{X}$ and $\mathbf{Y}$, albeit with a wider spread in $\mathbf{Y}$. This observation is supported by the histograms of these data shown in Figure 4.2. This data set is synthetic, so such observations may seem pointlessly obvious. However, if the data were not synthetic, then these observations would be useful to setting the tolerance domains, δ_x and δ_y, for the leaning. It has been shown that if A and B are known *a priori*, then the leaning will agree with intuition if $\delta_x = A$ and $\delta_Y = B$ [36]. It is assumed here, however, that A and B are unknown to the analyst. From Figure 4.2, initial tolerance domains of $\pm\delta_x = 0.5$ and $\pm\delta_y = 1$ will be used for the leaning calculations.

Autocorrelations in the data are useful to understand for both potential cause-effect assignments for the leaning calculations and potential issues in drawing causal inferences from the

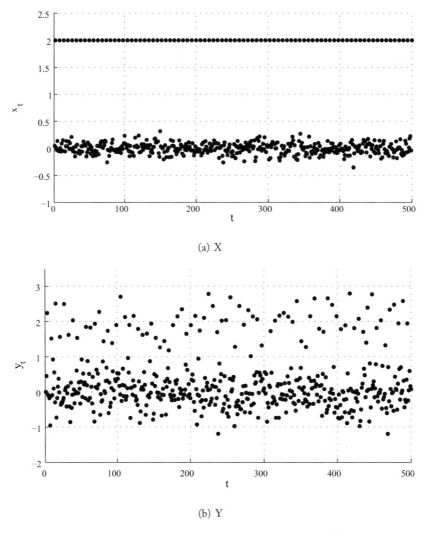

(a) X

(b) Y

Figure 4.1: An instance of Eq. (4.3) for $L = 500$, $A = 0.1$, and $B = 0.4$.

lagged cross-correlations. This data is synthetic, so it is known from Eq. (4.3) that a reasonable cause-effect assignment for the leaning would be $\{C, E\} = \{x_{t-1}, y_t\}$. The goal here is to show how such potential cause-effect assignments can be drawn directly from the data as part as ECA. Figure 4.3 shows strong autocorrelations for both **X** and **Y** at $l = 6, 12, 18, \ldots, 48$. The autocorrelations appear cyclic, which implies the leaning and lagged cross-correlation time series causality tools need not be calculated for more than $l = 1, 2, \ldots, 6$, which is the lag after which the autocorrelations pattern seems to repeat. If $l > 6$ in these calculations, then the causal inference may be strongly influenced by the autocorrelations in the data. The similar autocorrelations pattern

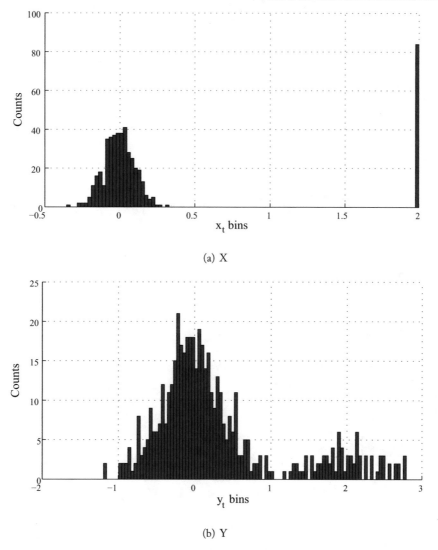

(a) X

(b) Y

Figure 4.2: Histograms of the instance of Eq. (4.3) shown in Figure 4.1.

seen in Figure 4.3 may imply a strong driving relationship within the time series pair (or perhaps a strong shared driving relationship with some outside driver) but do not immediately suggest a cause-effect assignment for the leaning calculation. The most straightforward cause-effect assignment is the l-standard assignment, which will be used with $l = 1, 2, \ldots, 6$.

The leaning calculations using the l-standard cause-effect assignment and tolerance domains suggested by Figures 4.3 and 4.2, respectively, fit naturally on a plot with the lagged

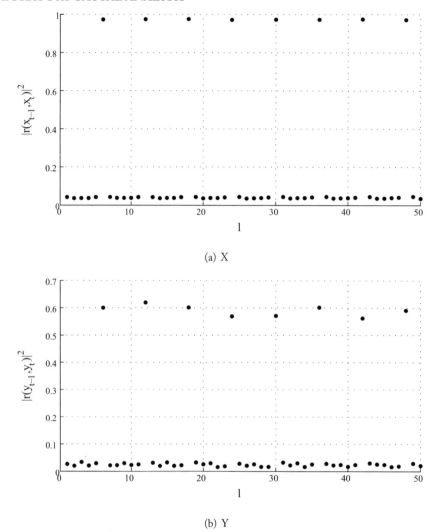

Figure 4.3: Autocorrelations of the instances of **X** and **Y** of Eq. (4.3) shown in Figure 4.1 given lags of $l = 1, 2, \ldots, 50$. The autocorrelations are $|r(b_{t-l}, b_t)|^2$ where $r(\cdot)$ is the Pearson correlation coefficient between the lagged l series $\{b_{t-l}\}$ and the time series $\{b_t\}$.

cross-correlations (because both depend on some lag l). These values are shown in Figure 4.4 for $l = 1, 2, \ldots, 6$. This figure shows both the weighted mean observed leaning, $\langle \lambda_l \rangle$, and the lagged cross-correlation differences, Δ_l, which suggest the same causal inference for almost every lag.[5] The causal inferences for the leanings are $\mathbf{Y} \to \mathbf{X}$ for $l = 2, \ldots, 6$ and $\mathbf{X} \to \mathbf{Y}$ for $l = 1$

---

[5]$\langle \lambda_l \rangle > 0$ and $\Delta_l < 0 \Rightarrow \mathbf{X} \to \mathbf{Y}$ and *vice versa*. See Sections 3.5 and 3.4.2 for more details.

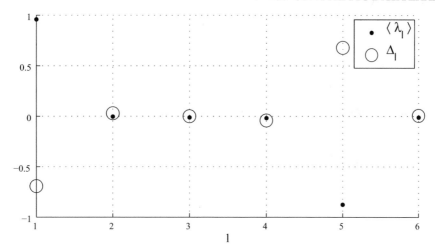

Figure 4.4: Lagged cross-correlation differences, Δ_l, and weighted mean observed leanings, $\langle \lambda_l \rangle$, (given an l-standard cause-effect assignment with $\pm \delta_x = 0.5$ and $\pm \delta_y = 1$) for the instances of Eq. (4.3) shown in Figure 4.1 given lags of $l = 1, 2, \ldots, 6$.

with a maximum absolute value at $l = 1$, which implies $\mathbf{X} \rightarrow \mathbf{Y}$, and a mean across all the lags, $\langle\langle \lambda_l \rangle\rangle_l = 6.6 \times 10^{-3}$, which also implies $\mathbf{X} \rightarrow \mathbf{Y}$. The causal inferences for the lagged cross-correlation differences are $\mathbf{Y} \rightarrow \mathbf{X}$ for $l = 2, 3, 5, 6$ and $\mathbf{X} \rightarrow \mathbf{Y}$ for $l = 1, 4$ with a maximum absolute value at $l = 1$, which implies $\mathbf{X} \rightarrow \mathbf{Y}$, and a mean across all the lags, $\langle \Delta_l \rangle_l = -2.8 \times 10^{-3}$, which also implies $\mathbf{X} \rightarrow \mathbf{Y}$. Both the mean across all lags and the maximum absolute value for both tools yield the same causal inference, and that causal inference agrees with intuition for this example. However, the second largest value of both tools in Figure 4.4 suggests the opposite (counter-intuitive) causal inference.

These lags can be investigated further by examining the correlation plots, i.e., plots of (y_{t-l}, x_t) (Figure 4.5) and (x_{t-l}, y_t) (Figure 4.6) for $l = 1, 2, \ldots, 6$.

These figures show the data clusters around 0 and 2, as expected from Figure 4.2. A notable feature is the strong linear relationship seen in Figure 4.6a and Figure 4.5e. These are the relationships shown by the lagged cross-correlation differences (and the leaning) in Figure 4.4. These relationships are not unexpected given the structure of Eq. (4.3). The (y_t, x_{t-1}) relationship is the explicit part of the system that accounts for, in large part, the intuitive causal structure. The (x_t, y_{t-5}) relationship is a by-product of the structured pulse pattern created for $\mathbf{X}$; i.e., if the pulses of $\mathbf{X}$ were set to occur at a different interval, then it is expected that this (x_t, y_{t-l}) relationship would appear for a lag l corresponding to that different interval.

The assumption, however, has been that the analyst doing the exploratory causal analysis of this synthetic data does not know Eq. (4.3) *a priori*. The question is what causal inference should such an analyst make for this example? It has already been shown that the causal inference

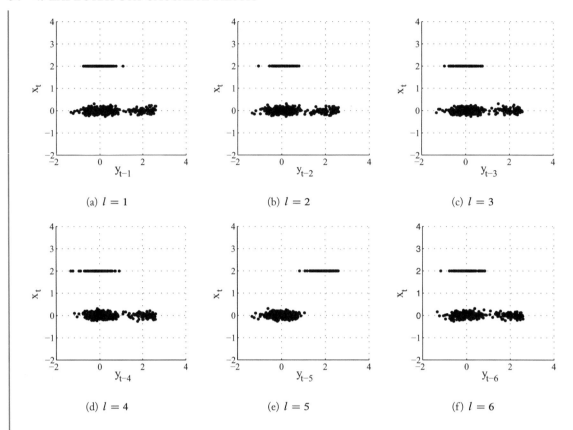

Figure 4.5: Lagged cross-correlation plots, (y_{t-l}, x_t), for the instance of Eq. (4.3) shown in Figure 4.1 given lags of $l = 1, 2, \ldots, 6$.

would be $\mathbf{X} \to \mathbf{Y}$ (which agrees with intuition) if the analyst were to use the average across all the calculated lags for either the leaning or the lagged cross-correlation difference, but the strong counter inferences provided by $l = 5$ (both $\langle \lambda_l \rangle$ and Δ_l imply $\mathbf{Y} \to \mathbf{X}$ for $l = 5$) may be seen as a concern that needs to be investigated further. The most desirable path forward may be to change the impulse pattern of $\mathbf{X}$ (e.g., by changing the period of the impulse) and see if the leaning and lagged cross-correlation differences change for either $l = 1$ (which implies the intuitive inference) or $l = 5$ (which implies the counter-intuitive inference). As explained in the previous paragraph, it is expected that doing so would change the result for $l = 5$ but may create a similar result for a different lag, while the $l = 1$ is expected to remain unchanged. This type of experimental result may be seen as strong evidence that the $l = 1$ results are the more correct causal inferences for this example system, but such experiments may not be possible. The analyst may only have access to the single instances of $\mathbf{X}$ and $\mathbf{Y}$ shown in Figure 4.1. Another possible approach is to calculate the leaning using other tolerance domains.

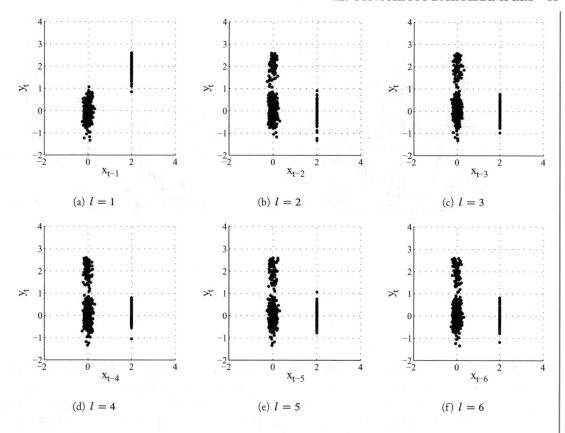

Figure 4.6: Lagged cross-correlation plots, (x_{t-l}, y_t), for the instance of Eq. (4.3) shown in Figure 4.1 given lags of $l = 1, 2, \ldots, 6$.

Consider the tolerance domains used in Figure 4.4, $\pm \delta_x = 0.5$ and $\pm \delta_y = 1$. These domains were found by visual inspection of Figures 4.1 and 4.2 (by estimating the widths of the peaks around 0). A computational approach is to set the tolerance domains as $\pm \delta_x = f(\max(\mathbf{X}) - \min(\mathbf{X}))$ and $\pm \delta_y = f(\max(\mathbf{Y}) - \min(\mathbf{Y}))$ with $f = 1/4$ (see [36] for other examples of setting tolerances domains). This method for setting the tolerance domains will be used often, so for brevity, it will be referred to as the *f-width* tolerance domains; i.e., this example uses the (1/4)-width tolerance domains. Figure 4.7 shows the leaning calculations using the l-standard assignment for $l = 1$ and $l = 5$ for different reasonable[6] tolerance domains of $\pm \delta_x \in [0, 1]$ and $\pm \delta_y \in [0, 1]$ in steps of 0.05. This figure shows the causal inferences implied for $l = 1$ and $l = 5$ for the method used in Figure 4.4 are consistently implied by the leaning calcula-

---

[6]"Reasonable" is defined in reference to Figure 4.2. For example, $\pm \delta_x = 2$ would lead to undefined leanings as the domain would be large enough to include all possible values of x_t and $\pm \delta_x = 1.5$ appears equivalent to $\pm \delta_x = 1.0$, so both such tolerance domains are considered "unreasonable."

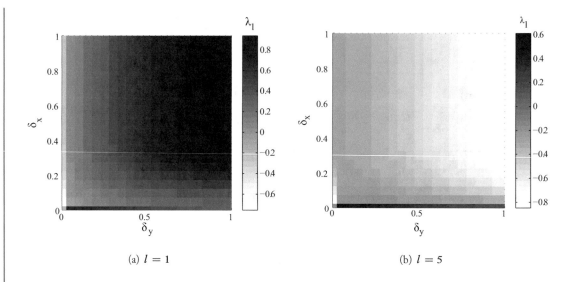

(a) $l = 1$ (b) $l = 5$

Figure 4.7: Leaning calculations for the instance of Eq. (4.3) shown in Figure 4.1 using the l-standard assignment for $l = 1$ and $l = 5$ and tolerance domains defined as $\pm\delta_x \in [0, 1]$ and $\pm\delta_y \in [0, 1]$ in steps of 0.05.

tions using different tolerance domains. Figure 4.7 shows the leaning calculation changes sign for small tolerance domains, as expected,[7] but is consistently either positive (for $l = 1$) or negative (for $l = 5$) as the tolerance domain increases. This behavior implies the signs of original leaning calculations were not attributable to the specific tolerance domains used in those calculations.

For this example, the MVGC toolbox returns Granger causality log-likelihood statistics of $F_{Y \to X} = 4.1 \times 10^{-3}$ and $F_{X \to Y} = 4.5 \times 10^{-1}$, or $F_{X \to Y} - F_{Y \to X} \approx 4.5 \times 10^{-1}$. The JIDT transfer entropy calculation returns $T_{X \to Y} = 6.0 \times 10^{-1}$ and $T_{Y \to X} = 6.9 \times 10^{-2}$, or $T_{X \to Y} - T_{Y \to X} = 5.3 \times 10^{-1}$. Both of these results imply $\mathbf{X} \to \mathbf{Y}$, which agrees with intuition. The PAI correlation difference is -8.3×10^{-3}, which also implies $\mathbf{X} \to \mathbf{Y}$. If the leaning and lagged cross-correlation difference contributions to the ECA summary vector are defined as the mean across all the tested lags, then the ECA summary for this example is $|\vec{g}|^2 = 0 \Rightarrow \mathbf{X} \to \mathbf{Y}$, which, again, is the intuitively correct answer.

Figure 4.8 shows the ECA summary for different instances of Eq. (4.3) evaluated with $A \in [0, 1]$ and $B \in [0, 1]$ in steps on 0.05 and $L = 500$. The leaning and lagged cross-correlation difference causal inferences were made with the average across all the tested lags, and the leaning calculations use the l-standard cause-effect assignment with $l = 1, 2, \ldots, 6$ (which are the same lags used by the lagged cross-correlation differences). The tolerance domains are calculated as $\pm\delta_x = f(\max(\mathbf{X}) - \min(\mathbf{X}))$ and $\pm\delta_y = f(\max(\mathbf{Y}) - \min(\mathbf{Y}))$ with $f = 1/4$. The maximum value of $|g|^2$ shown in Figure 4.8 is 1 which implies that the ECA summary never implies the

---

[7]See Section 3.5.

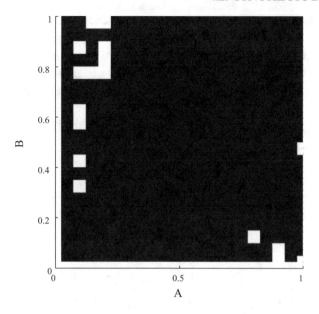

Figure 4.8: ECA summaries for different instances of Eq. (4.3) evaluated with $A \in [0, 1]$ and $B \in [0, 1]$ in steps on 0.05 and $L = 500$ with l-standard assignment leaning calculations and lagged cross-correlation differences given $l = 1, 2, \ldots, 6$ and leaning tolerance domains calculated as $\pm \delta_x = f(\max(\mathbf{X}) - \min(\mathbf{X}))$ and $\pm \delta_y = f(\max(\mathbf{Y}) - \min(\mathbf{Y}))$ with $f = 1/4$. The black indicates $|g|^2 = 0 \Rightarrow \mathbf{X} \to \mathbf{Y}$, as expected, and the white indicates $|g|^2 \neq 0$.

counter-intuitive causal inference of $\mathbf{Y} \to \mathbf{X}$ for this example. The majority of the elements of $\vec{g}$, i.e., g_1 (transfer entropy), g_2 (Granger), and g_4 (leaning), imply the intuitive causal inference of $\mathbf{X} \to \mathbf{Y}$ for almost every point[8] calculated in Figure 4.8. Counter-intuitive inferences are implied by g_3 (PAI) as A increases when B is relatively small (i.e., as the noise level in the impulse signal increases above the noise level in the response signal), and they are implied by g_5 (lagged cross-correlation) as B increases when A is relatively small (i.e., as the noise level in the response signal increases above the noise level in the impulse signal). Interestingly, both g_3 and g_5 imply the intuitive causal inference if $A \approx B$ for this example.

A more rigorous exploratory causal analysis of this example might investigate further those points in Figure 4.8 for which the ECA summary does not agree with intuition. It may be possible to change the PAI and lagged cross-correlation difference calculations, e.g., by using different embedding dimensions E and/or delay time steps τ in the PAI calculations or by pre-treating

---

[8]The causal inference for g_2 failed to be defined for approximately 20 points in Figure 4.8 because the MVGC toolbox failed to successfully fit a VAR model with the desired maximum model orders for some of the points with low noise in the impulse signal, e.g., $A = 0.05$. Every defined g_2, however, agreed with intuition, i.e., $g_2 = 0$. Every calculated g_1 and g_4 agreed with intuition.

the data in the lagged cross-correlation difference calculations. If, and how, such changes to the calculations change the causal inferences may provide insight into the relationship between the data sets. Such effort may seem inconsequential or trivial on a synthetic data set, but if an analyst has very little knowledge of the system dynamics that generated a given time series pair, then anything that may be drawn from exploratory causal analysis may be considered helpful.

4.2.2   CYCLIC DRIVING WITH LINEAR RESPONSE

Consider the linear example dynamical system of

$$\{\mathbf{X}, \mathbf{Y}\} = \{\{x_t\}, \{y_t\}\} \tag{4.4}$$

where $t = 0, 1, \ldots, L$,

$$x_t = a \sin(bt + c) + A\eta_t$$

and

$$y_t = x_{t-1} + B\eta_t$$

with $y_0 = 0$, $A \in [0, 1]$, $B \in [0, 1]$, $\eta_t \sim \mathcal{N}(0, 1)$, and with the amplitude a, the frequency b, and the phase c all in the appropriate units. This example is very similar to the previous one, except that the driving system $\mathbf{X}$ is sinusoidal.

Let the instance of Eq. (4.4) with $L = 500$, $A = 0.1$, $B = 0.4$, $a = b = 1$, and $c = 0$ shown in Figure 4.9 be considered the synthetic data set $(\mathbf{X}, \mathbf{Y})$ for the exploratory causal analysis. The histograms of these data are shown in Figure 4.10. These data have a wider spread than the data seen in Figure 4.1, and do not appear to be clustered about some small set of points. This observation is supported by Figure 4.10. The tolerance domains might be set by visual inspection of the histograms, as was done in Section 4.2.1, but for this example, the leaning calculation will use the $(1/4)$-width tolerance domains.

Figure 4.11 shows strong autocorrelations for both $\mathbf{X}$ and $\mathbf{Y}$, and the autocorrelations appear cyclic. So, the leaning and lagged cross-correlation time series causality tools will not be calculated for more than $l = 1, 2, \ldots, 20$, which is the lag after which the autocorrelations pattern seems to repeat. The most straightforward cause-effect assignment is the l-standard assignment, which will be used with $l = 1, 2, \ldots, 20$.

The leaning and lagged cross-correlation differences are shown in Figure 4.12 for $l = 1, 2, \ldots, 20$. This figure shows both the weighted mean observed leaning, $\langle \lambda_l \rangle$, and the lagged cross-correlation differences, Δ_l, seem to suggest different causal inferences for almost every third lag. The causal inference for both tools at $l = 1$ is the same, intuitively correct, inference of $\mathbf{X} \to \mathbf{Y}$, and the mean across all the lags is $\langle\langle \lambda_l \rangle\rangle_l = 3.9 \times 10^{-3}$ and $\langle \Delta_l \rangle_l = -2.9 \times 10^{-2}$, which also imply $\mathbf{X} \to \mathbf{Y}$.

For this example, the MVGC toolbox returns Granger causality log-likelihood statistics of $F_{Y \to X} = 2.8 \times 10^{-2}$ and $F_{X \to Y} = 2.4 \times 10^{-1}$, or $F_{X \to Y} - F_{Y \to X} = 2.1 \times 10^{-1}$. The JIDT transfer entropy calculation returns $T_{X \to Y} = 7.6 \times 10^{-1}$ and $T_{Y \to X} = 5.7 \times 10^{-1}$, or $T_{X \to Y} -$

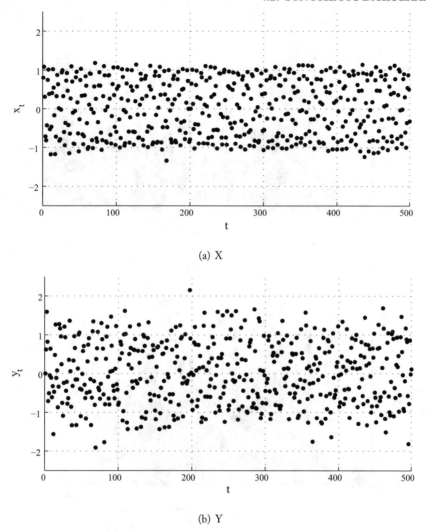

(a) X

(b) Y

Figure 4.9: An instance of Eq. (4.4) for $L = 500$, $A = 0.1$, and $B = 0.4$.

$T_{Y \to X} = 1.9 \times 10^{-1}$. The PAI correlation difference is -9.8×10^{-3}, which also implies $\mathbf{X} \to \mathbf{Y}$. If the leaning and lagged cross-correlation difference contributions to the ECA summary vector are defined as the mean across all the tested lags, then the ECA summary for this example is $|\vec{g}|^2 = 0 \Rightarrow \mathbf{X} \to \mathbf{Y}$. All of these results imply $\mathbf{X} \to \mathbf{Y}$, which agrees with intuition. Figure 4.13 shows the ECA summary for different instances of Eq. (4.4) evaluated with $A \in [0, 1]$ and $B \in [0, 1]$ in steps on 0.05 and $L = 250$. The leaning and lagged cross-correlation difference causal inferences were made with the average across all the tested lags, and the leaning calculations use

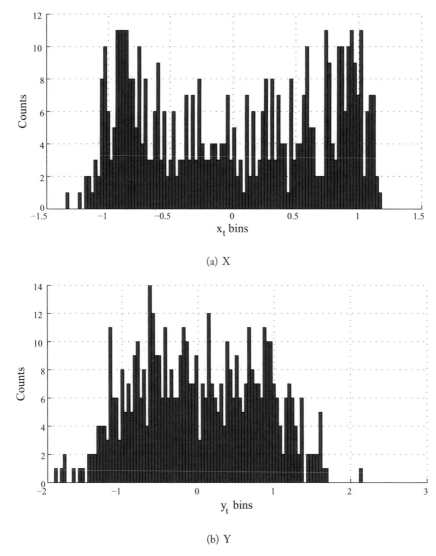

(a) X

(b) Y

Figure 4.10: Histograms of the instance of Eq. (4.4) shown in Figure 4.9.

the l-standard cause-effect assignment with $l = 1, 2, \ldots, 20$ (which are the same lags used by the lagged cross-correlation differences). The tolerance domains used the (1/4)-width domains. The maximum value of $|g|^2$ shown in Figure 4.13 is 1, not 5, which suggests that the ECA summary never implies the counter-intuitive causal inference of $\mathbf{Y} \rightarrow \mathbf{X}$ for this example. However, a visual comparison of Figures 4.8 and 4.13 shows the ECA summary agrees with intuition less reliably for this example than for the example of Section 4.2.1. The majority of the elements of $\vec{g}$, i.e.,

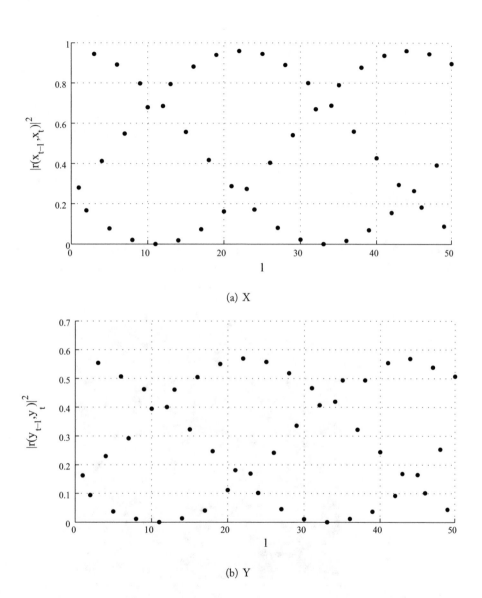

(a) X

(b) Y

Figure 4.11: Autocorrelations of the instances of **X** and **Y** of Eq. (4.4) shown in Figure 4.9 given lags of $l = 1, 2, \ldots, 50$. The autocorrelations are $|r(b_{t-l}, b_t)|^2$ where $r(\cdot)$ is the Pearson correlation coefficient between the lagged l series $\{b_{t-l}\}$ and the time series $\{b_t\}$.

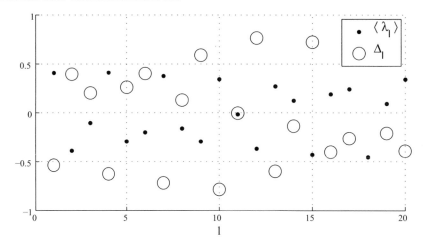

Figure 4.12: Lagged cross-correlation differences, Δ_l, and weighted mean observed leanings, $\langle \lambda_l \rangle$, (given an l-standard cause-effect assignment with $(1/4)$-width tolerance domains) for the instances of Eq. (4.4) shown in Figure 4.9 given lags of $l = 1, 2, \ldots, 20$.

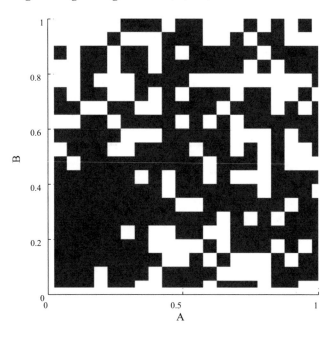

Figure 4.13: ECA summaries for different instances of Eq. (4.4) evaluated with $A \in [0, 1]$ and $B \in [0, 1]$ in steps on 0.05 and $L = 250$ with l-standard assignment leaning calculations and lagged cross-correlation differences given $l = 1, 2, \ldots, 20$ and $(1/4)$-width tolerance domains. The black indicates $|g|^2 = 0 \Rightarrow \mathbf{X} \to \mathbf{Y}$, as expected, and the white indicates $|g|^2 \neq 0$.

g_1 (transfer entropy), g_2 (Granger), and g_5 (lagged cross-correlation), imply the intuitive causal inference of $\mathbf{X} \rightarrow \mathbf{Y}$ for almost every point[9] calculated in Figure 4.13. The ECA summary failed to agree with intuition in this example for all the tested time series pairs because of counter-intuitive inferences implied by g_3 (PAI) and g_4 (leaning), which differs from the example shown in Section 4.2.1 where g_4 implied the intuitive causal inference for every tested time series pair. This observation helps illustrate the need for multiple types of tools in an exploratory causal analysis; the naive application of some tools may be better suited to a given system than others, and the use of different tools together can help guide the analyst in determining how to apply the time series causality tools differently (e.g., by changing the cause-effect assignment of the leaning or the model parameters of the Granger test statistic) for different systems. The inference implied by g_3 (PAI) is counter-intuitive as A increases, which is similar to the behavior seen for this tool in Section 4.2.1. The inference implied by g_4 (leaning) is counter-intuitive as B increases, which is in contrast to the behavior of g_3 (PAI) and implies the ECA summary may agree with intuition for most of the test times series pairs if either g_4 or g_3 were removed from $\vec{g}$.

As with the example in Section 4.2.1, a more rigorous exploratory causal analysis of this example might investigate further those points in Figure 4.13 for which the ECA summary does not agree with intuition. It may be possible to change the PAI and weighted mean observed leaning calculations, e.g., by using different embedding dimensions E and/or delay time steps τ in the PAI calculations or tolerance domains in the leaning calculations.

4.2.3   RL SERIES CIRCUIT

Consider a series circuit containing a resistor, inductor, and time varying voltage source related by

$$\frac{dI}{dt} = \frac{V(t)}{L} - \frac{R}{L}I(t), \tag{4.5}$$

where $I(t)$ is the current at time t, $V(t) = \sin(t)$ is the voltage at time t, R is the resistance, and L is the inductance. The time series pair for this example is then

$$\{\mathbf{V}, \mathbf{I}\} = \{\{V_t\}, \{I_t\}\} \tag{4.6}$$

where $\mathbf{V}$ is the set of discrete values of $V(t)$ evaluated using $t = 0, f\pi, 2f\pi, 3f\pi, \ldots, 8\pi$ with $f = 1/10$ and $\mathbf{I}$ is the set of discrete values found either by solving Eq. (4.5) numerically or by evaluating the analytical solution

$$I(t) = \frac{1}{D}\left(Le^{-\frac{t}{\tau}} + R\sin(t) - L\cos(t)\right) \tag{4.7}$$

with $D = L^2 + R^2$ and $\tau = L/R$, for the same time set used for $\mathbf{V}$.

---

[9]As with the example in Section 4.2.1, the MVGC toolbox failed to successfully fit a VAR model with the desired maximum model orders for some of the points with low noise in the impulse signal (three points were undefined). Again, every defined g_2 agreed with intuition, i.e., $g_2 = 0$. Every calculated g_1 and g_5 agreed with intuition.

Physical intuition is that V drives I, and so we expect to find that $\mathbf{V} \to \mathbf{I}$. These dynamics are well-known and have been thoroughly explored experimentally [216, 217]. As such, this example presents a slightly different test of the efficacy of the exploratory causal analysis than the last two examples. Previously, the intuitive causal inference relied on intuition about the mathematical form of the dynamics used to create the synthetic data. Here, the intuitive causal inference may be considered physical intuition and empirical experience, rather than any specific form or structure of Eq. (4.5).

Figure 4.14 shows $\mathbf{V}$ and $\mathbf{I}$ generated with both the analytical solution and a numerical solution to Eq. (4.5) using the *ode45* integration function in MATLAB. The time series $\mathbf{V}$ is created by defining values at fixed points and using linear interpolation to find the time steps required by the ODE solver. Two different physical scenarios are considered in which L and R are constant, $L = 10$ H and $R = 5$ Ω and $L = 5$ H and $R = 20$ Ω. Figure 4.14 shows the numerical and analytical solutions are very similar, but the causal inferences will be made using both for completeness. Figure 4.15 shows the 25-bin histograms for these signals.

Figure 4.16 shows, as expected, strong cyclic autocorrelations for both $\mathbf{V}$ and $\mathbf{I}$. The leaning and lagged cross-correlation differences are shown in Figure 4.17 for $l = 1, 2, \ldots, 5$. This figure shows both the weighted mean observed leaning, $\langle \lambda_l \rangle$, and the lagged cross-correlation differences, Δ_l, imply the intuitive causal inferences, i.e., $\mathbf{V} \to \mathbf{I}$, for every calculated lag. The mean across all the lags is $\langle \langle \lambda_l \rangle \rangle_l = 3.9 \times 10^{-1}$ and $\langle \Delta_l \rangle_l = -4.7 \times 10^{-1}$, for both the numerical and analytical solutions, given $L = 10$ H and $R = 5$ Ω, and $\langle \langle \lambda_l \rangle \rangle_l = 2.2 \times 10^{-1}$ and $\langle \Delta_l \rangle_l = -2.6 \times 10^{-1}$, for both the numerical and analytical solutions, given $L = 5$ H and $R = 20$ Ω. These values all imply $\mathbf{V} \to \mathbf{I}$.

The JIDT transfer entropy calculation returns $T_{V \to I} = 4.8 \times 10^{-1}$ and $T_{I \to V} = 4.6 \times 10^{-1}$, or $T_{V \to I} - T_{I \to V} = 1.7 \times 10^{-2}$ given $L = 10$ H and $R = 5$ Ω, and $T_{V \to I} = 3.5 \times 10^{-1}$ and $T_{I \to V} = 5.0 \times 10^{-2}$, or $T_{V \to I} - T_{I \to V} = 3.0 \times 10^{-1}$ given $L = 5$ H and $R = 20$ Ω. These values imply the intuitively correct causal inference, $\mathbf{V} \to \mathbf{I}$. The leaning and lagged cross-correlation difference contributions also imply the intuitive causal inference. The ECA summary for this example, however, is undefined for both physical scenarios because neither g_2 (Granger) nor g_3 (PAI) are defined. The PAI difference is undefined for these examples because the algorithm used to calculate the PAI correlations fails if two points in the shadow manifold have a separation distance of zero.[10] Likewise, the MVGC toolbox algorithms fail to return defined log-likelihood statistics for this example. These issues may be seen as technical issues regarding the different implementation algorithms or fundamental issues regarding the use of these techniques with "perfect" signals (such as $\mathbf{V}$ in this example). The use of multiple time series causality tools for the exploratory analysis, as has been done in this example, helps the analyst to understand the undefined results do not necessarily imply a lack of possible causal inferences.

---

[10]This scenario leads to a division by zero. See steps 2 and 3 of the algorithm outlined in Section 3.3. The implementation of the PAI algorithm used in this work declares the correlations undefined in such scenarios.

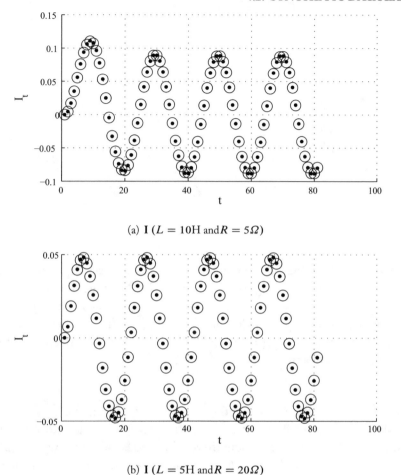

(a) $\mathbf{I}$ $(L = 10\text{H and} R = 5\Omega)$

(b) $\mathbf{I}$ $(L = 5\text{H and} R = 20\Omega)$

Figure 4.14: The voltage and current signals of a circuit containing a resistor R and inductor L in series where the driving voltage is $V(t) = \sin(t)$ and the current response is given by the solution to Eq. (4.5). Eq. (4.5) is solved both numerically (represented by the open dots in (a) and (b)) and analytically (represented by the solid dots in (a) and (b)). The signals are plotted for a sampling length of 8π seconds at a sampling interval of $10^{-1}\pi$ (or a sampling frequency of approximately 3 Hz), which corresponds to a time series length of 81 data points. *(Continues.)*

This example can also illustrate the importance of sample frequency and sample length. For example, the leaning calculation requires an assumed cause and effect pair to appear in the data enough times to provide reliable estimates of probabilities. Thus, data that is sampled for too few periods or too sparsely may lead to leanings that do not agree with intuition. Consider the signals shown in Figure 4.14 with $L = 10$ H and $R = 5$ Ω sampled at the time steps $t =$

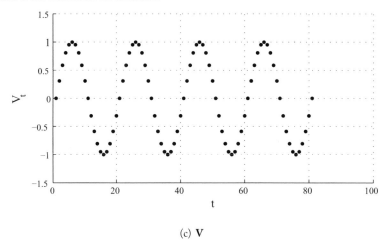

(c) **V**

Figure 4.14: *(Continued.)* The voltage and current signals of a circuit containing a resistor R and inductor L in series where the driving voltage is $V(t) = \sin(t)$ and the current response is given by the solution to Eq. (4.5). Eq. (4.5) is solved both numerically (represented by the open dots in (a) and (b)) and analytically (represented by the solid dots in (a) and (b)). The signals are plotted for a sampling length of 8π seconds at a sampling interval of $10^{-1}\pi$ (or a sampling frequency of approximately 3 Hz), which corresponds to a time series length of 81 data points.

$0, f\pi, 2f\pi, 3f\pi, \ldots, 8\pi$ for $f = 80^{-1}, 60^{-1}, 40^{-1}, 20^{-1}, 10^{-1}$ and 5^{-1}. The tolerance domains for the leaning calculation will be set as the $(1/4)$-width domains, and the number of lags to average over for the ECA summary using the leaning and lagged cross-correlation difference will be set as $l = 1, \ldots, l_a$, where l_a is the lag for which the autocorrelation of the voltage signal is minimum in a set of autocorrelations calculated with lags from 1 to half the length of **V**. This method is not equivalent to the method used to produce Figure 4.17 (which relied on a visual inspection of Figure 4.16), but it is easier to implement computationally and for $f = 10^{-1}$ it provides leaning and lagged cross-correlation difference calculations that imply the same causal inference as Figure 4.17. Table 4.1 shows the variation of $\vec{g}$ as f changes. The ECA summary is undefined at each sampling frequency but g_3 (PAI) begins to imply the intuitive causal inference as f in increases. Nothing about the signals has physically changed but the times series data (upon which the causal inference is drawn) has changed because of the changing sampling procedure.

4.2.4   CYCLIC DRIVING WITH NON-LINEAR RESPONSE

Consider the nonlinear dynamical system of

$$\{\mathbf{X}, \mathbf{Y}\} = \{\{x_t\}, \{y_t\}\} \tag{4.8}$$

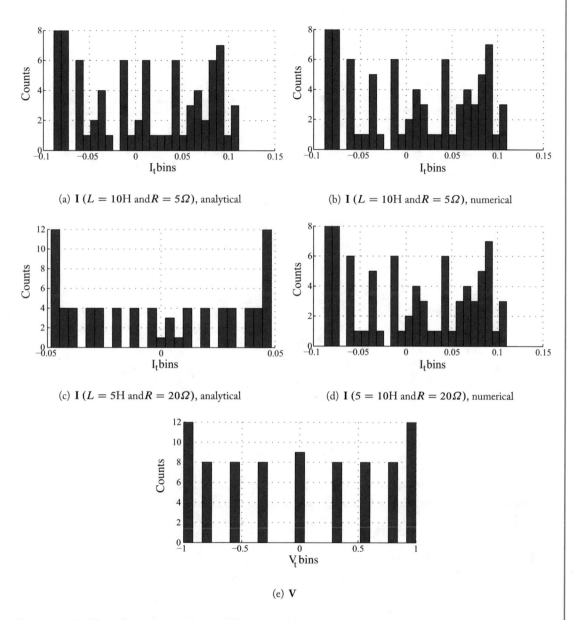

(a) $\mathbf{I}$ ($L = 10$H and $R = 5\Omega$), analytical

(b) $\mathbf{I}$ ($L = 10$H and $R = 5\Omega$), numerical

(c) $\mathbf{I}$ ($L = 5$H and $R = 20\Omega$), analytical

(d) $\mathbf{I}$ ($5 = 10$H and $R = 20\Omega$), numerical

(e) $\mathbf{V}$

Figure 4.15: The 25-bin histograms of the signals shown in Figure 4.14.

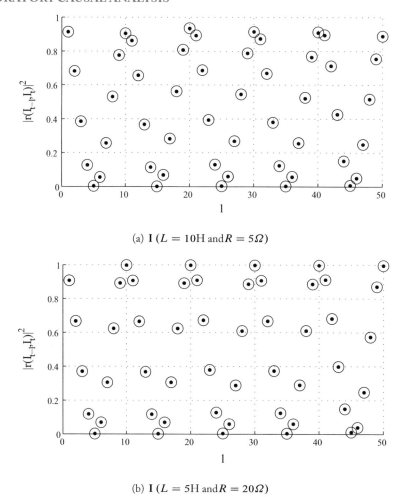

(a) I $(L = 10\text{H and} R = 5\Omega)$

(b) I $(L = 5\text{H and} R = 20\Omega)$

Figure 4.16: Autocorrelations of the signals shown in Figure 4.14. The numerical solution for **I** is represented by the open dots in (a) and (b) and the analytical solution is represented by the solid dots in (a) and (b). *(Continues.)*

where $t = 0, 1, \ldots, L$,

$$x_t = a \sin(bt + c) + A\eta_t$$

and

$$y_t = B x_{t-1} (1 - C x_{t-1}) + D\eta_t,$$

with $y_0 = 0$, with $A, B, C, D \in [0, 1]$, $\eta_t \sim \mathcal{N}(0, 1)$, and with the amplitude a, the frequency b, and the phase c all in the appropriate units given $t = 0, f\pi, 2f\pi, 3f\pi, \ldots, 6\pi$ with $f = 1/30$, which implies $L = 181$.

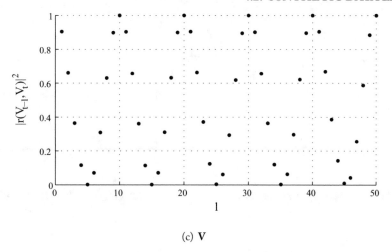

(c) **V**

Figure 4.16: *(Continued.)* Autocorrelations of the signals shown in Figure 4.14. The numerical solution for **I** is represented by the open dots in (a) and (b) and the analytical solution is represented by the solid dots in (a) and (b).

Table 4.1: The ECA summary and vector, $\vec{g}$, for the signals shown in Figure 4.14 with $L = 10$ H and $R = 5$ Ω sampled at the time steps $t = 0, f\pi, 2f\pi, 3f\pi, \ldots, 8\pi$

f	$\vec{g}$	ECA summary
80^{-1}	$(0, 2, 2, 0, 0)$	undefined
60^{-1}	$(0, 2, 2, 0, 0)$	undefined
40^{-1}	$(0, 2, 2, 0, 0)$	undefined
20^{-1}	$(0, 2, 2, 0, 0)$	undefined
10^{-1}	$(0, 2, 2, 0, 0)$	undefined
5^{-1}	$(0, 2, 2, 0, 0)$	undefined

Let the instance of Eq. (4.8) with $A = 0.1$, $B = 0.3$, $C = 0.4$, $D = 0.5$, $a = b = 1$, and $c = 0$ shown in Figure 4.18 be the time series pair $(\mathbf{X}, \mathbf{Y})$ for the exploratory causal analysis. The tolerance domains for the leaning calculation are set as the $(1/4)$-width tolerance domains.

Figure 4.19 shows strong cyclic autocorrelations for **X**. The autocorrelations of **Y** are apparently acyclic, which might make it difficult to set the number of leaning and lagged cross-correlation difference lags to calculate based on **Y** in Figure 4.19. The leaning (using the l-standard assignment) and lagged cross-correlation time series causality tools will not be calculated for more than $l = 1, 2, \ldots, 15$, which is the lag after which the autocorrelations pattern seems to repeat for **X**.

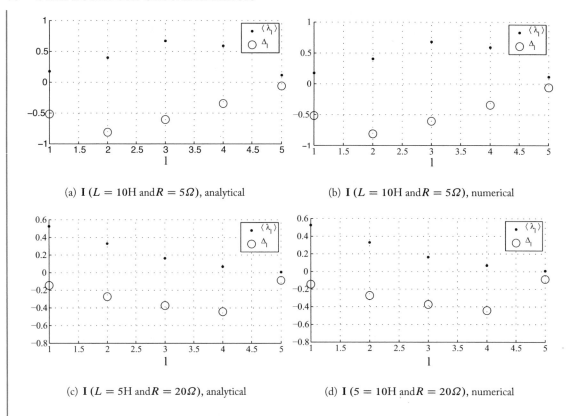

(a) $\mathbf{I}$ ($L = 10\mathrm{H}$ and $R = 5\Omega$), analytical

(b) $\mathbf{I}$ ($L = 10\mathrm{H}$ and $R = 5\Omega$), numerical

(c) $\mathbf{I}$ ($L = 5\mathrm{H}$ and $R = 20\Omega$), analytical

(d) $\mathbf{I}$ ($5 = 10\mathrm{H}$ and $R = 20\Omega$), numerical

Figure 4.17: Lagged cross-correlation differences, Δ_l, and weighted mean observed leanings, $\langle \lambda_l \rangle$, (given an l-standard cause-effect assignment with $(1/4)$-width tolerance domains) for the signals shown in Figure 4.14 given lags of $l = 1, 2, \ldots, 5$.

The leaning and lagged cross-correlation differences are shown in Figure 4.20 for $l = 1, 2, \ldots, 20$. This figure shows both the weighted mean observed leaning, $\langle \lambda_l \rangle$, and the lagged cross-correlation differences, Δ_l, seem to suggest the intuitive causal inference, $\mathbf{X} \rightarrow \mathbf{Y}$, for the majority of the calculated lags. The mean across all the lags is $\langle \langle \lambda_l \rangle \rangle_l = 8.4 \times 10^{-3}$ and $\langle \Delta_l \rangle_l = -6.8 \times 10^{-2}$, which both imply $\mathbf{X} \rightarrow \mathbf{Y}$.

For this example, the MVGC toolbox returns Granger causality log-likelihood statistics of $F_{X \rightarrow Y} - F_{Y \rightarrow X} = 2.6 \times 10^{-1}$. The JIDT transfer entropy calculation returns $T_{X \rightarrow Y} - T_{Y \rightarrow X} = 2.7 \times 10^{-1}$. The PAI correlation difference is -1.8×10^{-3}, which also implies $\mathbf{X} \rightarrow \mathbf{Y}$. If the leaning and lagged cross-correlation difference contributions to the ECA summary vector are defined as the mean across all the tested lags, then the ECA summary for this example is $|\vec{g}|^2 = 0 \Rightarrow \mathbf{X} \rightarrow \mathbf{Y}$. All of these results imply $\mathbf{X} \rightarrow \mathbf{Y}$, which agrees with intuition.

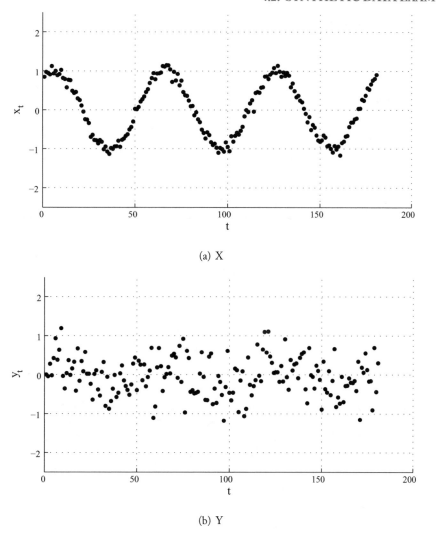

(a) X

(b) Y

Figure 4.18: An instance of Eq. (4.8) for $A = 0.1$, $B = 0.3$, $C = 0.4$, $D = 0.5$, $a = b = 1$, and $c = 0$.

A natural question is whether the data series shown in Figure 4.18 is a unique instance of Eq. (4.8). The synthetic data examples have, so far, interpreted the instances of Eq. (4.3), (4.4), and (4.8) being explored as the only collection of data points potentially collected by an analyst. However, the stochastic noise present in each of these examples raises the question of whether or not the ECA summaries for these examples would have agreed with intuition for a different instance of these systems; i.e., for instances of Eq. (4.3), (4.4), and (4.8) with the same coefficients

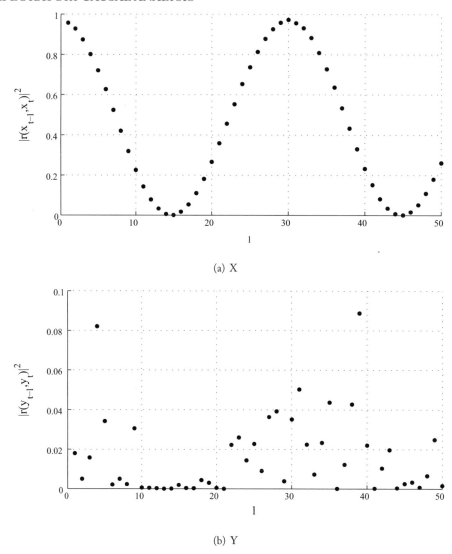

(a) X

(b) Y

Figure 4.19: Autocorrelations of the instances of **X** and **Y** of Eq. (4.8) shown in Figure 4.18 given lags of $l = 1, 2, \ldots, 50$. The autocorrelations are $|r(b_{t-l}, b_t)|^2$ where $r(\cdot)$ is the Pearson correlation coefficient between the lagged l series $\{b_{t-l}\}$ and the time series $\{b_t\}$.

as those used to create Figures 4.1, 4.9, and 4.18 but different realizations of the stochastic noise terms η_t. In many physical systems, an analyst may be able to measure several instances of **X** and **Y**. Exploratory causal analysis may be performed on the collection of times series pairs. For example, given m sets of the pair $(\mathbf{X}, \mathbf{Y})$, then an analyst may combine the time series in some way

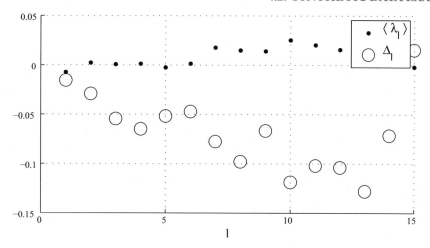

Figure 4.20: Lagged cross-correlation differences, Δ_l, and weighted mean observed leanings, $\langle\lambda_l\rangle$, (given an l-standard cause-effect assignment with $(1/4)$-width tolerance domains) for the instances of Eq. (4.8) shown in Figure 4.18 given lags of $l = 1, 2, \ldots, 15$.

(e.g., $\bar{\mathbf{X}} = \{\bar{x}_t = m^{-1}\sum_i^m x_{t,i} \; \forall t = 1, 2, \ldots, L\}$ where $\mathbf{X}_i = \{x_{t,i}\}$ is the ith measured instance of $\mathbf{X}$, which contains L data points) and form an ECA summary used the combined data series, or an analyst may calculate m ECA summaries which may then be used for causal inference.

Consider 10^4 instances of Eq. (4.8) with the coefficients shown in Figure 4.18. Table 4.2 shows the ECA summary vector distribution for this collection of time series pairs. The individ-

Table 4.2: The ECA summary distribution of 10^4 instances of Eq. (4.8) with the coefficients shown in Figure 4.18

$\vec{g}$	Counts	ECA summary
$(0, 0, 0, 0, 0)$	5707	undefined
$(0, 0, 0, 0, 1)$	231	undefined
$(0, 0, 0, 1, 0)$	1608	undefined
$(0, 0, 0, 1, 1)$	2454	undefined

ual causal inferences g_1 (transfer entropy), g_2 (Granger),[11] and g_3 (PAI) agree with intuition for every instance. The two that imply counterintuitive causal inferences do so only for a minority of the tested instances, approximately 41% of the instances for g_4 (leaning) and approximately

---

[11]It is interesting to note that the MVGC toolbox uses a linear VAR model fitting procedure that implies the intuitively correct causal inference for every instance of this nonlinear example. This might be seen as evidence for using every available time series causality tool during exploratory causal inference, even those with well documented shortcomings.

27% of the instances for g_5 (lagged cross-correlation). The leaning calculation uses $(1/4)$-width tolerance domains and both the leaning and the lagged cross-correlation differences are calculated as the mean of lags $l = 1, 2, \ldots, 15$. The average leaning, $\langle\langle\lambda_l\rangle\rangle$, across all 10^4 instances is $\langle\langle\lambda_l\rangle\rangle = 4 \times 10^{-3}$ and the average lagged cross-correlation difference, $\langle\Delta_l\rangle = -4.3 \times 10^{-2}$, both of which imply the intuitive causal inference. A bootstrapping [218] procedure can be set up with the sample of leaning and lagged cross-correlation calculations, whereby 10^6 means are calculated from new sets (of the same size as the original set) of leanings and lagged cross-correlations that have been sampled (with replacement) from the original set. This procedure yields no negative means for the leaning calculation and no positive means for the lagged cross-correlation differences; the null hypothesis that the mean leaning value is negative (i.e., $\langle\langle\lambda_1^z\rangle\rangle < 0$) and the null hypothesis that the mean lagged cross-correlation difference is positive (i.e., $\langle\Delta_l\rangle > 0$) can be rejected with a p-value less than 10^{-6}. The 90% confidence interval for the mean of the 10^6 bootstrapped leaning calculation means is $[3.7 \times 10^{-3}, 4.3 \times 10^{-3}]$ and for the lagged cross-correlation means is $[-4.4 \times 10^{-2}, -4.1 \times 10^{-2}]$. These results imply the intuitively correct causal inference for both g_4 and g_5.

It becomes more difficult to visualize the agreement between the ECA summary and intuition for different sets of system parameters as the parameter space becomes larger. Eq. (4.8) has a 4-dimensional parameter space in which such agreement may be explored, A, B, C, and D where a, b, and c are held constant. One option for visualization may be plotting points within the unit cube framed by B, C, and D for a given A where $|g|^2 = 0$. Every point tested within the 4-D parameter space for which the ECA summary agrees with intuition should appear in one of the units cubes. The number of plots, however, becomes unwieldy if A is sampled for more than a few points on the unit domain, and the reader would rely on counting the number of points on each plot and comparing it to the total number of tested points to determine how often the ECA summary failed to agree with intuition. Rather than produce such a set of plots, the 4-D parameter space will be sampled to provide descriptive statistics. The four parameters were sampled in steps of 0.05 in the domains of $A \in [0.05, 0.55]$ and $B, C, D \in [0.05, 1.0]$ for a total of 88,000 instances of Eq. (4.8). The ECA summary agreed with intuition, i.e., $|\vec{g}|^2 = 0$, in 60,155 (68%) of those instances. Most of the time series causality tools that failed to imply the intuitive causal inference did so by implying the counter-intuitive causal inference, i.e., $g_i \neq 2$ for any $i \neq 2$ of any of the ECA summary vectors. The only exception was g_2 (Granger) which failed to provide any causal inference if a VAR forecast model could not be fit to the data within the requested maximum model order by the MVGC toolbox. The fewest counter-intuitive inferences were implied by g_1 (transfer entropy) with 381 or 0.43% of the total number of tested instances. The majority (327 or 86%) of those 381 counter-intuitive implications occurred for instances with $C > 0.5$ and/or $D > 0.5$. The majority (3,414 or 73%) of the 4,686 (5.3% of the total number of tested instances) counter-intuitive inferences implied by g_3 (PAI) occurred for instances with $B > 0.5$ and/or $C > 0.5$. The highest number of counter-intuitive inferences were implied by g_4 (leaning) and g_5 (lagged cross-correlation) with 19,971 and 14,051, or 23% and 16% of the total number of instances, respectively. The counter-intuitive inferences implied by both tools occurred mostly (65% for g_4 and 68% for g_5) for instances with $D > 0.5$. The number of lags used in the calculation of both the leaning and the lagged cross-correlation difference was fixed (with $l = 1, 2, \ldots, 15$) for each tested instance of Eq. (4.8). More intuitive causal inferences may have been implied by g_4 and g_5 if the number of lags used in those calculations was algorithmically set, e.g., with autocorrelation lengths as was discussed in Section 4.2.3. The (1/4)-width tolerance domains used in the leaning calculation may also have contributed to the counter-intuitive inferences in some instances. This assumption might be checked by varying the tolerance domains in the leaning calculation for every tested instance of Eq. (4.8) for which the (1/4)-width tolerance domain calculations implied a counter-intuitive causal inference.

4.2.5   COUPLED LOGISTIC MAP

Consider the nonlinear dynamical system of

$$\{\mathbf{X}, \mathbf{Y}\} = \{\{x_t\}, \{y_t\}\} \tag{4.9}$$

where $t = 0, 1, \ldots, L$,

$$x_t = x_{t-1}\left(r_x - r_x x_{t-1} - \beta_{xy} y_{t-1}\right)$$

and

$$y_t = y_{t-1}\left(r_y - r_y y_{t-1} - \beta_{yx} x_{t-1}\right)$$

where the parameters $r_x, r_y, \beta_{xy}, \beta_{yx} \in \mathbb{R} \geq 0$. This pair of equations is a specific form of the two-dimensional coupled logistic map system often used to model population dynamics [219] and it was a system used in the introduction of cross convergent mapping, CCM, which is a SSR time series causality tool [30].

Sugihara et al. [30] note that $\beta_{xy} > \beta_{yx}$ intuitively implies $\mathbf{Y}$ "drives" $\mathbf{X}$ more than $\mathbf{X}$ "drives" $\mathbf{Y}$, and vice versa. Such intuition, however, can be difficult to justify for all instances of Eq. (4.9). The x_{t-1} term that appears in y_t can be seen as a function of x_{t-2} with coefficients of $\beta_{yx} r_x$. These product coefficients suggest that if $r_x > r_y$, then $\mathbf{X}$ may be seen as the stronger driver in the system even if $\beta_{yx} < \beta_{xy}$. The same argument can be made, with the appropriate substitutions, to show that $\mathbf{Y}$ may be seen as the stronger driver in the system even if $\beta_{xy} < \beta_{yx}$. As such, there is no clear intuitive causal inference for this system.

Consider the instance of Eq. (4.9) with $L = 500$, $\beta_{xy} = 0.5$, $\beta_{yx} = 1.5$, $r_x = 3.8$, and $r_y = 3.2$ with initial conditions $x_0 = y_0 = 0.4$ shown in Figure 4.21. For this example, $r_x > r_y$, $\beta_{yx} > \beta_{xy}$, and the initial conditions are the same, which implies the intuitive causal inference for this example is $\mathbf{X} \rightarrow \mathbf{Y}$. The histograms of these data are shown in Figure 4.22. For this example, the leaning calculation will use the (1/4)-width tolerance domains.

Figure 4.23 shows strong autocorrelations for both $\mathbf{X}$ and $\mathbf{Y}$ for a lag of $l = 1$ and progressively weaker autocorrelations for $l = 2, 3, 4$ with minimal autocorrelations for $l \geq 5$. The autocorrelations do not appear cyclic. The argument was made in Section 4.2.1 that cyclic patterns in the autocorrelations of either time series might be used to limit the number of lags for

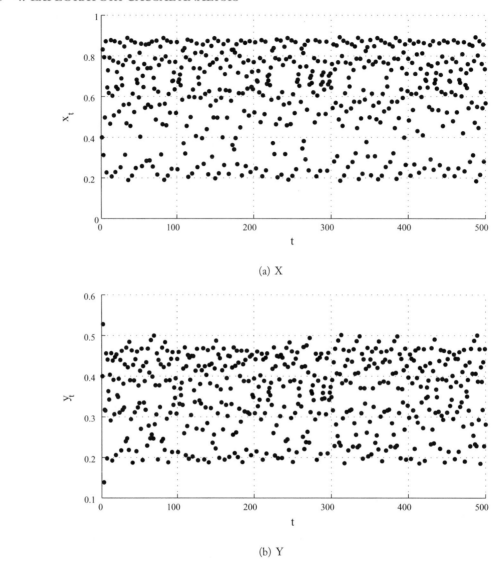

Figure 4.21: An instance of Eq. (4.9) for $L = 500$, $\beta_{xy} = 0.5$, $\beta_{yx} = 1.5$, $r_x = 3.8$, and $r_y = 3.2$ with initial conditions $x_0 = y_0 = 0.4$.

which the leanings and lagged cross-correlation differences are calculated, i.e., it was argued that a repeating autocorrelation pattern after a given lag l may imply that the leaning or cross-correlation calculations using lags greater than l would give redundant information for drawing causal inferences. This argument is not applicable for this example given that there are no apparent cyclic autocorrelation patterns. There is no *a priori* reason to assume reasonable cause-effect assignments

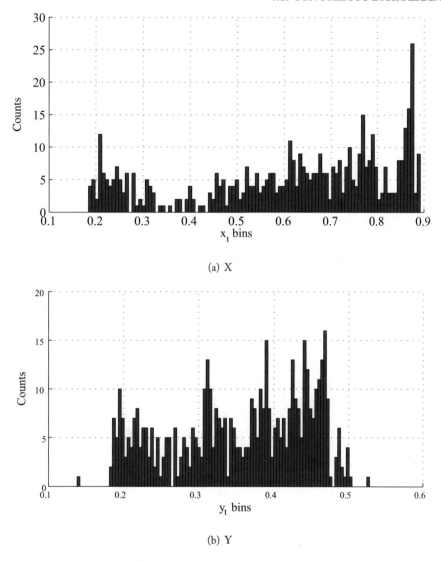

(a) X

(b) Y

Figure 4.22: Histograms of the instance of Eq. (4.9) shown in Figure 4.21.

for the leaning might be drawn from autocorrelation patterns (i.e., the relationship of a signal to itself at different points in time is not necessarily related to the relationship between that signal and its potential driving or response partner signal). However, visual inspection of the signals in Figure 4.21 shows the signals have (roughly) similar shapes. So, it may be reasonable to assume that if the signal **X** has a strong relationship with its own past at a given l, then, by a loose similarity argument, the partner signal **Y** may also have a strong relationship with **X** at lag l. For this

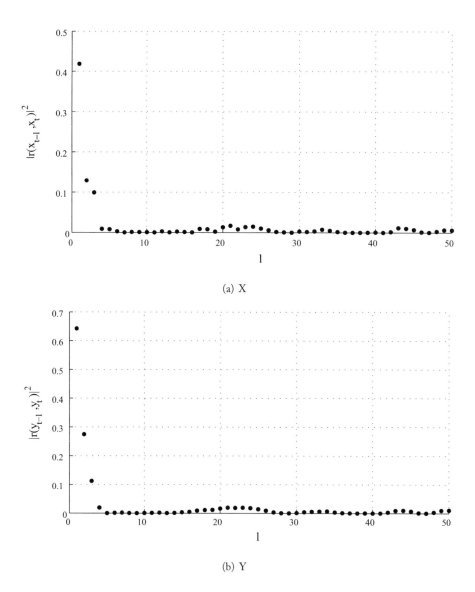

(a) X

(b) Y

Figure 4.23: Autocorrelations of the instances of **X** and **Y** of Eq. (4.9) shown in Figure 4.21 given lags of $l = 1, 2, \ldots, 50$. The autocorrelations are $|r(b_{t-l}, b_t)|^2$ where $r(\cdot)$ is the Pearson correlation coefficient between the lagged l series $\{b_{t-l}\}$ and the time series $\{b_t\}$.

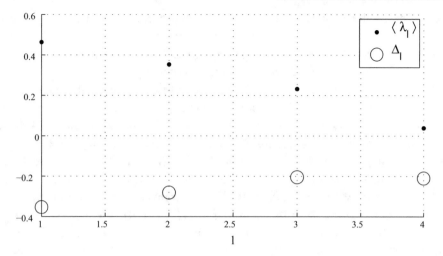

Figure 4.24: Lagged cross-correlation differences, Δ_l, and weighted mean observed leanings, $\langle \lambda_l \rangle$, (given an l-standard cause-effect assignment with $(1/4)$-width tolerance domains) for the instances of Eq. (4.4) shown in Figure 4.21 given lags of $l = 1, 2, \ldots, 4$.

example, the leaning (with the l-standard assignment) and lagged cross-correlation time series causality tools will be calculated for lags $l = 1, 2, \ldots, 4$, given the autocorrelations seem to decrease significantly for $l \geq 5$. A cause-effect assignment must be made to calculate the leaning. It will be shown that the leaning calculated using the l-standard assignment with $= 1, 2, \ldots, 4$ implies the same causal inference as the other time series causality tools used in this example.

The leaning and lagged cross-correlation differences are shown in Figure 4.24 for $l = 1, 2, \ldots, 4$. This figure shows both the weighted mean observed leaning, $\langle \lambda_l \rangle$, and the lagged cross-correlation differences, Δ_l, imply the intuitively correct causal inference $\mathbf{X} \rightarrow \mathbf{Y}$. The mean across all the lags is $\langle \langle \lambda_l \rangle \rangle_l = 2.7 \times 10^{-1}$ and $\langle \Delta_l \rangle_l = -2.6 \times 10^{-1}$, both of which imply $\mathbf{X} \rightarrow \mathbf{Y}$.

For this example, the MVGC toolbox returns Granger causality log-likelihood statistics of $F_{X \rightarrow Y} - F_{Y \rightarrow X} = 5.4 \times 10^{-1}$. The JIDT transfer entropy calculation returns $T_{X \rightarrow Y} - T_{Y \rightarrow X} = 4.9 \times 10^{-1}$. The PAI correlation difference is -3.9×10^{-3}. If the leaning and lagged cross-correlation difference contributions to the ECA summary vector are defined as the mean across all the tested lags, then the ECA summary for this example is $|\vec{g}|^2 = 0 \Rightarrow \mathbf{X} \rightarrow \mathbf{Y}$. All of these results imply $\mathbf{X} \rightarrow \mathbf{Y}$, which agrees with intuition.

The parameter space of Eq. (4.9) provides an opportunity to ask many interesting questions about the exploratory causal analysis. For example, an instance of Eq. (4.9) with $L = 500$, $\beta_{xy} = \beta_{yx} = 1.0$, $r_x = r_y = 3.5$ and $x_0 = y_0 = 0.4$ leads to an ECA summary vector of $\vec{g} = (2, 2, 2, 2, 2)$ (given g_4 [leaning] and g_5 [lagged cross-correlation] calculated in the same fashion as the previous example, i.e., Figure 4.18). The intuitive causal inference for this instance of Eq. (4.8) is not obvious and the undefined ECA summary seems to imply this confusion is not remedied by the exploratory causal analysis approach. Changing the system parameters slightly to $L = 500$, $\beta_{xy} = \beta_{yx} = 1.0$, $r_x = 3.6$, $r_y = 3.4$ and $x_0 = y_0 = 0.4$ leads to an ECA summary vector of $\vec{g} = (1, 2, 1, 0, 0)$. It may be argued that the intuitive causal inference in this case is $\mathbf{X} \rightarrow \mathbf{Y}$ because $r_x > r_y$ with all the other systems parameters being equal. This intuition, however, is not reflected by the ECA summary vector, where only two of the time series causality tools imply that intuitive inference, g_4 (leaning) and g_5 (lagged cross-correlation). Consider instead $L = 500$, $\beta_{xy} = 1.1$, $\beta_{yx} = 0.9$, $r_x = r_y = 3.5$ and $x_0 = y_0 = 0.4$, which leads to an ECA summary vector of $\vec{g} = (0, 2, 2, 0, 1)$. The intuitive causal inference in this case might be $\mathbf{Y} \rightarrow \mathbf{X}$ because $\beta_{xy} > \beta_{yx}$ with all the other systems parameters being equal, but only one tool, g_5 (lagged cross-correlation), implies this inference in the ECA summary vector. The initial conditions of the system also affect the ECA summary, e.g., $L = 500$, $\beta_{xy} = \beta_{yx} = 1.0$, $r_x = r_y = 3.5$, $x_0 = 0.3$ and $y_0 = 0.5$ leads to $\vec{g} = (1, 2, 2, 1, 0)$. The ECA summary does, however, seem to agree with intuition when such intuitions are straightforward for a given instance of Eq. (4.8); e.g., $L = 500$, $\beta_{xy} = \beta_{yx} = 0.5$, $r_x = 3.0$, $r_y = 3.8$, and $x_0 = y_0 = 0.4$ leads to $\vec{g} = (1, 1, 0, 1, 1)$ (which implies $\mathbf{Y} \rightarrow \mathbf{X}$, as expected, for all g_i except g_3 (PAI)) and $L = 500$, $\beta_{xy} = 0.5$, $\beta_{yx} = 2.0$, $r_x = r_y = 3.5$, and $x_0 = y_0 = 0.4$ leads to $\vec{g} = (0, 2, 2, 0, 0)$ (which implies $\mathbf{X} \rightarrow \mathbf{Y}$, as expected, for a majority of the g_i).

4.2.6   IMPULSE WITH MULTIPLE LINEAR RESPONSES

Consider the multivariate system of

$$\{\mathbf{X}, \mathbf{Y}, \mathbf{Z}\} = \{\{x_t\}, \{y_t\}, \{z_t\}\} \tag{4.10}$$

where $t = 0, 1, \ldots, L$,

$$x_t = \begin{cases} 2 & t = 1 \\ A\eta_t & \forall\, t \in \{t \mid t \neq 1 \text{ and } t \bmod 5 \neq 0\} \\ 2 & \forall\, t \in \{t \mid t \bmod 5 = 0\} \end{cases}$$

and

$$y_t = x_{t-1} + B\eta_t \ ,$$

and either (case 1)

$$z_t = y_{t-1} \tag{4.11}$$

or (case 2)

$$z'_t = y_{t-1} + y_t = y_{t-1} + x_{t-1} + B\eta_t \tag{4.12}$$

or (case 3)

$$z''_t = y_{t-1} + x_{t-1} + z''_{t-1} \tag{4.13}$$

with $y_0 = 0$, $B \in \mathbb{R} \geq 0$, $\eta_t \sim \mathcal{N}(0, 1)$, and $L = 500$.

In case 1, **Z** depends directly on **Y** and indirectly on **X** (through **Y**, which depends directly on **X**). The intuitive causal inference is then **Y** → **Z** and **X** → **Z**. Case 2, despite the additional **Y** dependence in **Z**, has the same intuitive causal inference as case 1. In case 3, **Z** depends directly on itself and both **Y** and **X**. Case 3 also has the same intuitive causal inference.

Consider the instance of Eq. (4.3) with $L = 500$, $A = 0.4$, and $B = 0.6$ shown in Figures 4.25 and 4.26. The leaning calculation will use (1/4)-width tolerance domains.

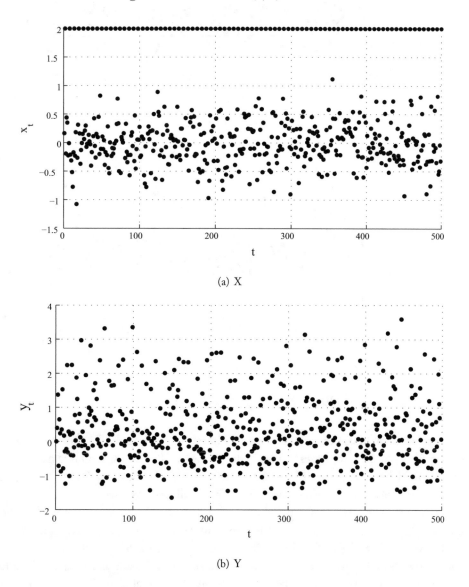

(a) X

(b) Y

Figure 4.25: An instance of Eq. (4.10) (**X** and **Y**) for $L = 500$, $A = 0.4$, and $B = 0.6$.

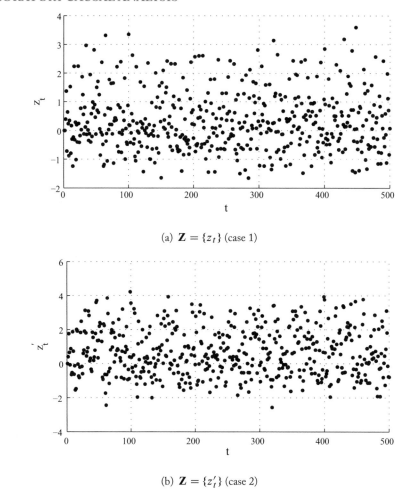

(a) $\mathbf{Z} = \{z_t\}$ (case 1)

(b) $\mathbf{Z} = \{z_t'\}$ (case 2)

Figure 4.26: An instance of Eq. (4.10) $(\mathbf{Z})$for $L = 500$, $A = 0.4$, and $B = 0.6$. *(Continues.)*

Histograms of these data are shown in Figure 4.27. Figure 4.28 shows strong autocorrelations for $\mathbf{X}$, $\mathbf{Y}$, and $\mathbf{Z}$ (for cases 1 and 2) at $l = 6, 12, 18, \ldots, 48$, which is expected given the similar forms of Eq. (4.10) and (4.3) (see Section 4.2.1). The leaning (using the l-standard assignment) and lagged cross-correlation calculations will use lags $l = 1, 2, \ldots, 6$.

Figure 4.29 shows both the weighted mean observed leaning, $\langle \lambda_l \rangle$, and the lagged cross-correlation differences, Δ_l, for each of the seven time series pairs in this example, $(\mathbf{X}, \mathbf{Y})$, $(\mathbf{X}, \mathbf{Z})$ (for all three cases), and $(\mathbf{Y}, \mathbf{Z})$ (for all three cases). Table 4.3 shows the values of each of the five time series causality tools used in this exploratory causal analysis for each of the time series pairs in this example with the mean leaning across all the lags labeled "L," the mean lagged cross-correlation difference labeled "LCC," Granger causality log-likelihood statistics difference (i.e.,

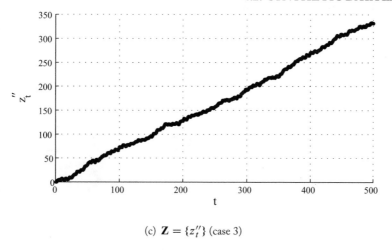

(c) $\mathbf{Z} = \{z_t''\}$ (case 3)

Figure 4.26: *(Continued.)* An instance of Eq. (4.10) $(\mathbf{Z})$for $L = 500$, $A = 0.4$, and $B = 0.6$.

$F_{X \to Y} - F_{Y \to X}$, calculated by the MVGC toolbox) labeled "GC," the transfer entropy difference (i.e., $T_{X \to Y} - T_{Y \to X}$, calculated with the JIDT) labeled "TE," and the PAI correlation difference labeled "PAI." Table 4.3 also shows the ECA summary and vector for each pair.

Table 4.3 shows the individual values for each tool used during the exploratory analysis along with the ECA summary vector. Table 4.4 shows how these ECA summaries compare both the intuitive causal inferences for each case and to the majority-vote inference, i.e, the causal inference implied by the majority of the time series causality tools used during the analysis. The ECA summary of $\mathbf{X} \to \mathbf{Y}$, and its agreement with the intuitive inference, may have been expected given the similarity of this example to the one discussed in Section 4.2.1, but the only two other ECA summaries that agree with the intuitive inferences are for case 1 and 2 of the pair $(\mathbf{X}, \mathbf{Z})$. The ECA summary is undefined for every other time series pair. The ECA summary was undefined for case 1 and 2 of the pair $(\mathbf{Y}, \mathbf{Z})$ because g_2 (Granger) was undefined. In these scenarios, along with case 3 of both pairs, the MVGC toolbox failed to fit a VAR model to the data with the maximum request model parameters and/or within the maximum allotted computation time. However, the majority-vote inference agrees with intuition for case 1 and 2 of the pair $(\mathbf{Y}, \mathbf{Z})$. Interestingly, the majority-vote inference for case 3 of both time series pairs is counter-intuitive. Only g_4 (leaning) agrees with intuition for the case 3 pairs. These counter-intuitive majority-vote inferences may imply case 3 of $\mathbf{Z}$ has some property that makes time series causality particularly unreliable. Case 3 of $\mathbf{Z}$ is, e.g., apparently non-stationary.[12] The autoregressive term in z_t'' of Eq. (4.10) is unique among all the cases of $\mathbf{Z}$ in this example. These properties may lead to unreliable exploratory causal analysis with the time series causality tools being used in this work. This idea might be

---

[12]The phrase "apparently" is used here to indicate that the approximate stationarity of the data is drawn from visual inspections of Figures 4.25 and 4.26 rather than formal tests.

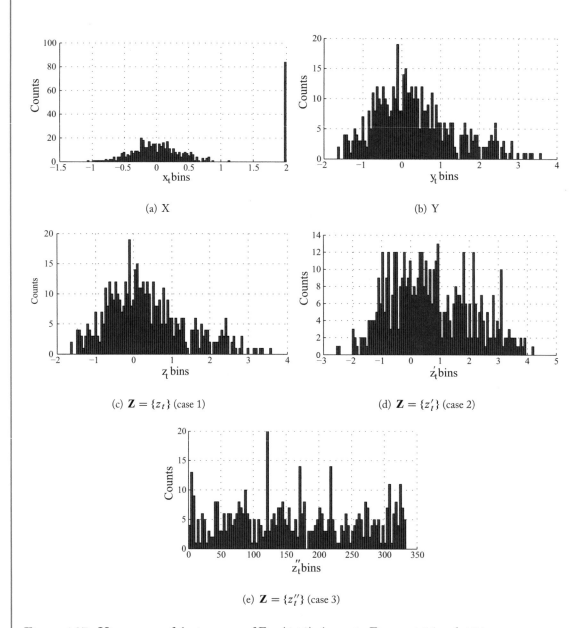

(a) X

(b) Y

(c) $\mathbf{Z} = \{z_t\}$ (case 1)

(d) $\mathbf{Z} = \{z'_t\}$ (case 2)

(e) $\mathbf{Z} = \{z''_t\}$ (case 3)

Figure 4.27: Histograms of the instance of Eq. (4.10) shown in Figures 4.25 and 4.26.

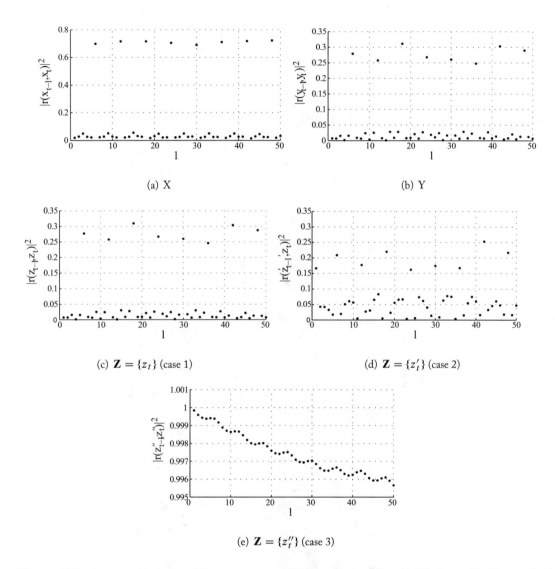

Figure 4.28: Autocorrelations of the instances of $\mathbf{X}$, $\mathbf{Y}$, and $\mathbf{Z}$ of Eq. (4.10) shown in Figures 4.25 and 4.26 given lags of $l = 1, 2, \ldots, 50$. The autocorrelations are $|r(b_{t-l}, b_t)|^2$ where $r(\cdot)$ is the Pearson correlation coefficient between the lagged l series $\{b_{t-l}\}$ and the time series $\{b_t\}$.

Table 4.3: The values of each of the five time series causality tools used in this exploratory causal analysis for each of the time series pairs shown in Figures 4.25 and 4.26 with the mean leaning across all the lags labeled "L," the mean lagged cross-correlation difference labeled "LCC," Granger causality log-likelihood statistics difference (i.e., $F_{X \to Y} - F_{Y \to X}$, calculated by the MVGC toolbox) labeled "GC," the transfer entropy difference (i.e., $T_{X \to Y} - T_{Y \to X}$, calculated with the JIDT) labeled "TE," and the PAI correlation difference labeled "PAI." An entry of "n/a" in the GC column indicates the MVGC toolbox failed to fit a VAR model to the data with the maximum request model parameters and/or within the maximum allotted computation time.

	TE	GC	PAI	L	LCC	$\vec{g}$	ECA summary
(X,Y)	5.9×10^{-1}	8.0×10^{-1}	-6.7×10^{-3}	2.3×10^{-2}	-1.4×10^{-2}	$(0, 0, 0, 0, 0)$	$X \to Y$
(X,Z) (case 1)	2.7×10^{-2}	7.9×10^{-1}	-1.2×10^{-2}	2.2×10^{-2}	-1.6×10^{-2}	$(0, 0, 0, 0, 0)$	$X \to Z$
(X,Z) (case 2)	2.8×10^{-1}	6.5×10^{-1}	-6.8×10^{-3}	2.5×10^{-2}	-3.7×10^{-2}	$(0, 0, 0, 0, 0)$	$X \to Z$
(X,Z) (case 3)	-2.7×10^{-2}	n/a	4.1×10^{-3}	5.9×10^{-4}	1.9×10^{-3}	$(1, 2, 1, 0, 01)$	undefined
(Y,Z) (case 1)	1.8	n/a	-5.4×10^{-3}	1.2×10^{-1}	-7.7×10^{-2}	$(0, 2, 0, 0, 0)$	undefined
(Y,Z) (case 2)	3.2×10^{-1}	n/a	-6.9×10^{-5}	3.6×10^{-2}	-6.0×10^{-2}	$(0, 2, 0, 0, 0)$	undefined
(Y,Z) (case 3)	-4.6×10^{-2}	n/a	1.1×10^{-2}	4.1×10^{-4}	7.5×10^{-3}	$(1, 2, 1, 0, 01)$	undefined

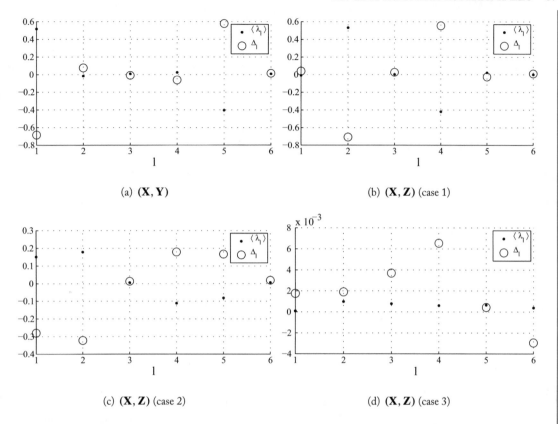

Figure 4.29: Lagged cross-correlation differences, Δ_l, and weighted mean observed leanings, $\langle\lambda_l\rangle$, (given an l-standard cause-effect assignment with $(1/4)$-width tolerance domains) for the instances of Eq. (4.10) shown in Figures 4.25 and 4.26 given lags of $l = 1, 2, \ldots, 6$. *(Continues.)*

explored by focusing on more synthetic data examples with response signals similar to case 3 of **Z**, i.e., with autoregressive terms and/or non-stationary. In this example, however, the counter-intuitive inferences illustrate that exploratory causal analysis inferences may be unreliable in the presence of confounding. Although, if an analyst only had access to **Y** and **Z** in this example, then the majority-vote inferences would still imply the intuitive causal inference in every case except case 3, as seen in Table 4.4.

4.3   EMPIRICAL DATA EXAMPLES

Empirical data sets with known (or assumed) causal relationships may be used to understand how exploratory causal inference using leanings might be done if the system dynamics are unknown (or sufficiently complicated to make first principle numerical comparisons cumbersome).

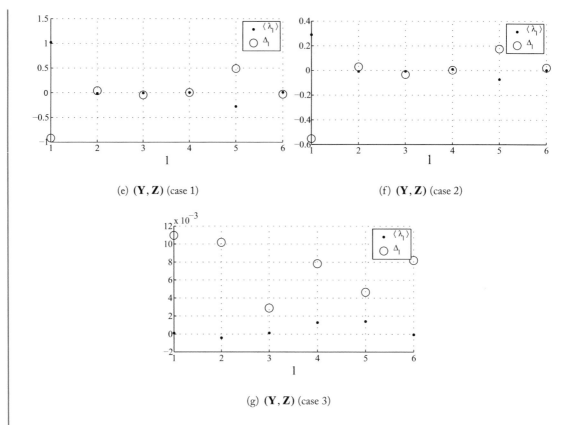

(e) **(Y, Z)** (case 1)   (f) **(Y, Z)** (case 2)

(g) **(Y, Z)** (case 3)

Figure 4.29: *(Continued.)* Lagged cross-correlation differences, Δ_l, and weighted mean observed leanings, $\langle \lambda_l \rangle$, (given an l-standard cause-effect assignment with (1/4)-width tolerance domains) for the instances of Eq. (4.10) shown in Figures 4.25 and 4.26 given lags of $l = 1, 2, \ldots, 6$.

4.3.1   SNOWFALL DATA

Figure 4.30 shows a time series pair with causal "truth" from the UCI Machine Learning Repository (MLR) [220]. This data repository is a collection of data sets (some of which are time series) with known, intuitive, or assumed causal relationships meant for use in the testing of causal discovery algorithms in machine learning [220].

Figure 4.30a,b are times series of the daily snowfall (the expected response) and mean temperature (the expected driver) from July 1, 1972, to December 31, 2009, at Whistler, BC, Canada (Latitude: 50°04′04.000″ N, Longitude: 122°56′50.000″ W, Elevation: 1835.00 meters). From [220], "Common sense tells us that X [mean temperature] causes Y [snow fall] (with maybe very small feedback of Y on X). Confounders are present (e.g., day of the year)." These time series correspond to data set 87 of the MLR [220]. Thus, following the notation of Figure 4.30, the intuitive causal inference for this example is $\mathbf{X} \to \mathbf{Y}$.

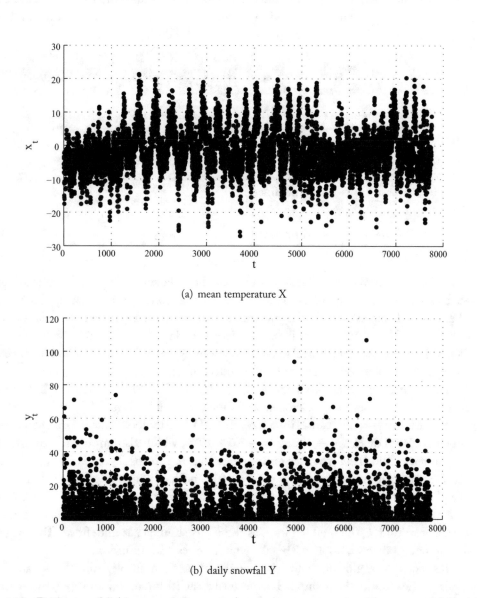

(a) mean temperature X

(b) daily snowfall Y

Figure 4.30: Daily snowfall (the expected response) and mean temperature (the expected driver) from July 1, 1972, to December 31, 2009, at Whistler, BC, Canada (Latitude: 50°04′04.000″ N, Longitude: 122°56′50.000″ W, Elevation: 1835.00 meters).

Table 4.4: Comparisons of the ECA summaries for each of the time series pairs shown in Figures 4.25 and 4.26 with the intuitive inferences and the "majority-vote" inference, i.e., the causal inference implied by the majority of the times series causality tools used during the exploratory causal analysis

	ECA summary	Majority-vote inference	Intuitive inference
(X,Y)	X→Y	X→Y	X→Y
(X,Z) (case 1)	X→Z	X→Z	X→Z
(X,Z) (case 2)	X→Z	X→Z	X→Z
(X,Z) (case 3)	undefined	Z→X	X→Z
(Y,Z) (case 1)	undefined	Y→Z	Y→Z
(Y,Z) (case 2)	undefined	Y→Z	Y→Z
(Y,Z) (case 3)	undefined	Z→Y	Y→Z

Figure 4.32 shows the autocorrelations for the data shown Figure 4.30, and Figure 4.31 shows the 100-bin histograms of the data. The autocorrelations do not appear cyclic (within the 50 lags that were calculated) but, for the snowfall time series (i.e., **Y**), the autocorrelations approach zero for $l > 20$. This observation will be used to set the lags for the leaning (with the l-standard assignment) and lagged cross-correlation calculations as $l = 1, 2, \ldots, 20$. The tolerance domains for the leaning calculation will be the $(1/4)$-width domains.

The mean leaning across all the lags, $\langle\langle \lambda_l \rangle\rangle_l = 3.7 \times 10^{-2}$, which implies the intuitive causal inference. The mean lagged cross-correlation across all the lags, $\langle \Delta_l \rangle_l = 2.3 \times 10^{-2}$, which implies the counter-intuitive causal inference. The MVGC toolbox returns Granger causality log-likelihood statistics of $F_{X \to Y} - F_{Y \to X} = -2.6 \times 10^{-3}$, which also implies the counter-intuitive causal inference. The JIDT transfer entropy calculation returns $T_{X \to Y} - T_{Y \to X} = 2.1 \times 10^{-2}$, and the PAI correlation difference[13] is -3.4×10^{-2}, both of which imply the intuitive causal inference. If the leaning and lagged cross-correlation difference contributions to the ECA summary vector are defined as the mean across all the tested lags, then the ECA summary vector for this example is $\vec{g} = (0, 1, 0, 0, 1)$, which implies the ECA summary is undefined. The majority-vote causal inference, however, implies $\mathbf{X} \to \mathbf{Y}$, which agrees with intuition.

This example again illustrates the benefit of using multiple time series causality tools. Consider g_4 (leaning), which implied an intuitive causal inference, and g_5 (cross-correlation), which implied a counter-intuitive causal inference. If those tools are calculated as the means across all tested lags with $l = 1, 2, \ldots, l_{max}$, then as seen above, $l_{max} = 20 \Rightarrow g_4 = 0$, $g_5 = 1$.

---

[13]For this example, the embedding dimension was set as $E = 100$, rather than $E = 3$, which was the embedding dimension in every previous example. The larger embedding dimension was not set with any formal procedure (see, e.g., [159–161, 221, 222]). Instead, the PAI correlation difference was calculated with $E = 100$ once it was discovered that the algorithm failed with $E = 3$ (See footnote in Section 4.2.3 for a discussion of failures of the PAI algorithm). The time delay used in this example was the same as every other example, i.e., $\tau = 1$.

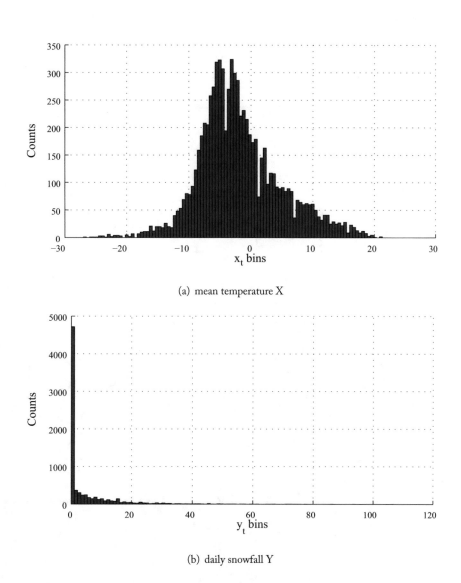

(a) mean temperature X

(b) daily snowfall Y

Figure 4.31: Histograms of the data shown in Figure 4.30.

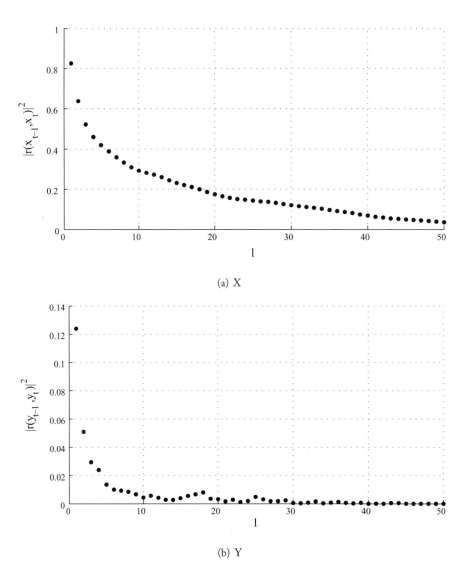

(a) X

(b) Y

Figure 4.32: Autocorrelations of the data shown in Figure 4.30 given lags of $l = 1, 2, \ldots, 50$. The autocorrelations are $|r(b_{t-l}, b_t)|^2$ where $r(\cdot)$ is the Pearson correlation coefficient between the lagged l series $\{b_{t-l}\}$ and the time series $\{b_t\}$.

It is also true that any $l_{max} = 2, 3, \ldots, 20 \Rightarrow g_4 = 0$, $g_5 = 1$. The lagged cross-correlation difference with the maximum absolute value across all the tested lags is positive, which also implies the counter-intuitive inference. The majority of the lagged cross-correlation differences implied the counter-intuitive inference (16 of the 20 calculated lags). So, it seems most reasonable approaches to determine g_5 from the set of lagged cross-correlation differences would lead to $g_5 = 1$, which does not agree with intuition. Likewise, g_2 (Granger) implies the counter-intuitive causal inference; i.e., $g_2 = 1$. It may be argued that the MVGC log-likelihood of $F_{X \to Y} - F_{Y \to X} = -2.6 \times 10^{-3}$ is better interpreted as $g_2 = 2$ because the value is two orders of magnitude closer to zero than any other MVGC log-likelihood differences calculated in this work. But, neither conclusion, $g_2 = 1$ or $g_2 = 2$, agrees with the intuitive causal inference. The use of either of these time series causality tools alone (i.e., g_2 or g_5) may lead an analyst to draw a counter-intuitive causal inference. The use of these tools as part of a set, however, allows the analyst to compare and contrast the different tools. The analyst may decide to use a majority-vote inference, which agrees with intuition for this example, or an analyst may use outside assumptions to determine certain tools, e.g., g_5 or g_2, are unreliable for the type of data being analyzed. At a minimum, the different implications of the tools strongly suggests the analyst should consider applying each tool more carefully, e.g., by changing the model parameters for g_2.

4.3.2   OMNI DATA

The NASA OMNI data set consists of hourly-averaged time series measurements of several different space weather parameters from 1963 to present, collected from more than twenty different satellites, along with sunspot number and several different geomagnetic indices, including D_{st}, collected from the NOAA National Geophysical Data Center [14]. The disturbance storm time, D_{st}, is a measure of geomagnetic activity [223]. The magnetic field measurements in the OMNI data sets, specifically B_z in GSE coordinates [224], is believed to be a driver of D_{st} [225].

Consider the available measurements for the magnetic field, B_z (in nanoTesla, nT), and the disturbance storm time, D_{st}. The collection and aggregation procedures used to create the OMNI data set can be found in [14].[14] The time series pair $(\mathbf{B}, \mathbf{D})$ consists of all the available measurements $\mathbf{B} = \{B_z(t)\}$ and $\mathbf{D} = \{D_{st}(t)\}$ where $\Delta t = t_{n+1} - t_n = 1$ hour $\forall n = 1, 2, \ldots, N - 1$ (with $N = 438312$), the first time step $t_0 = 0$ represents 00:30:00 January 1, 1963 UTC, and the last time step $t_N = 438312$ represents 23:30:00 December 31, 2012 UTC. These hourly data sets are shown in Figure 4.33. The 1000-bin histograms of these times series are shown in Figure 4.34. It is standard practice to "rectify" the magnetic field component B_z as B_s where $B_s(t) = B_z(t)$ if $B_z(t) \leq 0$ and $B_s(t) = 0$ if $B_z(t) > 0$ [226, 227]. This treatment leads to a second times series pair $(\mathbf{B}_r, \mathbf{D})$ with $\mathbf{B}_r = \{B_s(t)\}$. Figure 4.35 shows both the times series and 1000-bin histogram of $\mathbf{B}_r$.

The time series $\mathbf{B}$, $\mathbf{B}_r$, and $\mathbf{D}$ are long, with approximately 440,000 data points each, and each contains missing data (i.e., time stamps for which no measurements are present in the data

---

[14]The data itself, along with discussions of format, units, and origin, can also be found at http://omniweb.gsfc.nasa.gov/.

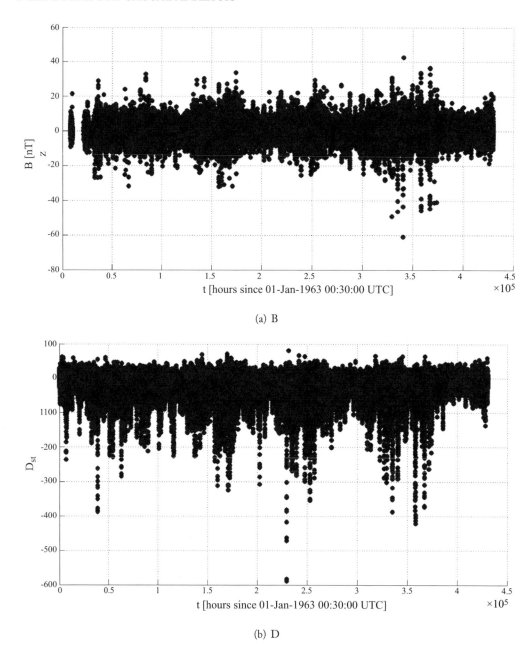

(a) B

(b) D

Figure 4.33: Hourly measurements of the magnetic field component, B_z, and disturbance storm time index, D_{st} from the beginning of January 1, 1963, until the end of December 31, 2012, taken from the NASA OMNI data set.

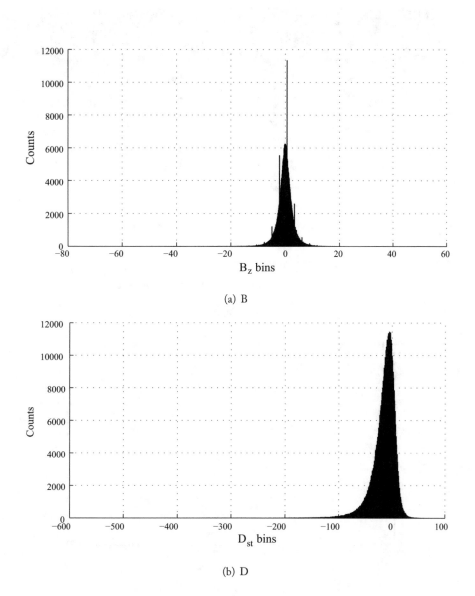

(a) B

(b) D

Figure 4.34: Histograms of the data shown in Figure 4.33.

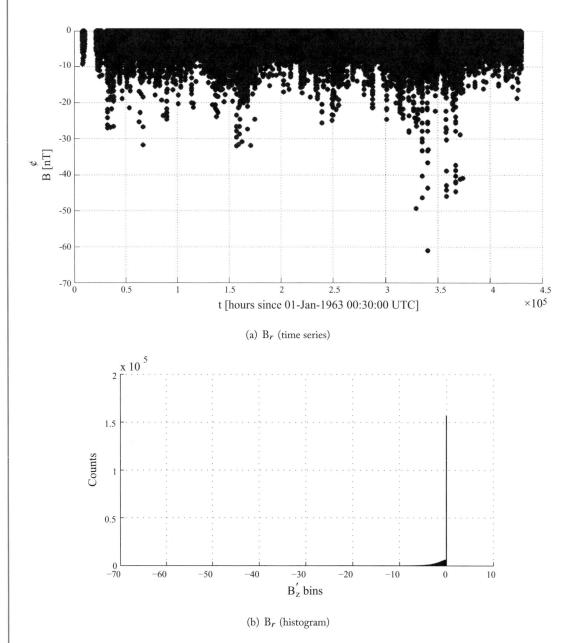

(a) B_r (time series)

(b) B_r (histogram)

Figure 4.35: Hourly measurements of the rectified magnetic field, B_s from the beginning of January 1, 1963, until the end of December 31, 2012, and the 1000-bin histogram of that data.

set). Both of these properties are practical concerns for the data analyst. The length may make the calculation of certain time series causality tools computationally infeasible, and the missing data may lead to erroneous results, if the algorithm used to calculate a given time series causality tool can even handle missing data. The magnetic field data, $\mathbf{B}$, contains 131,928 missing data points, and the storm index, D_{st}, contains 7,736 missing data points. Some, but not all, of the missing data coincide within the time series pair. Creating a new time series from the defined points of $\mathbf{B}$ and $\mathbf{D}$ will lead to nonuniform time intervals for subsequent time steps in the series, which will make causal interpretations more difficult.[15] The length of the time series may be addressed by only considering contiguous subsets of the series, but there is a risk that data associated with interesting physical phenomena captured during one subset may not be present in another. This work will address these practical issues with two different approaches, averaging and sampling.

There is no missing data in $\mathbf{D}$ between 00:30:00 January 1, 1997 UTC and 23:30:00 December 31, 2001 UTC. Let $\bar{\mathbf{D}}$, $\bar{\mathbf{B}}$, and $\bar{\mathbf{B}}_r$ be the daily (i.e., 24 hour) averages of $\mathbf{D}$, $\mathbf{B}$, and $\mathbf{B}_r$, respectively, within this time period, where a daily average is the arithmetic mean of 24 hours of data ignoring any missing values. Each of the new time series, $\bar{\mathbf{D}}$, $\bar{\mathbf{B}}$, and $\bar{\mathbf{B}}_r$, contains 1,826 data points with no missing points. Figures 4.36, 4.37, and 4.38 show both the times series and the 100-bin histograms of these averaged data sets.

The autocorrelations of $\bar{\mathbf{D}}$ and $\bar{\mathbf{B}}$ are shown in Figure 4.39, from which the lags for the leaning (using the l-standard cause-effect assignment) and lagged cross-correlation difference calculations are set as $l = 1, 2, \ldots, 30$. The leaning calculation will use the $(1/4)$ tolerance domains. The PAI correlation differences are calculated with $E = 100$ and $\tau = 1$. ECA summary vectors can be constructed for the two time series pairs $(\bar{\mathbf{B}}, \bar{\mathbf{D}})$ and $(\bar{\mathbf{B}}_r, \bar{\mathbf{D}})$, which are shown in Tables 4.6 and 4.5. The majority-vote inference agrees with intuition for both pairs, but the ECA summary only agrees with intuition for the pair $(\bar{\mathbf{B}}_r, \bar{\mathbf{D}})$. The ECA summary for the pair $(\bar{\mathbf{B}}, \bar{\mathbf{D}})$ is undefined because g_2 (Granger) implies $\bar{\mathbf{B}} \leftarrow \bar{\mathbf{D}}$ while every other time series causality tool implies the intuitively correct $\bar{\mathbf{B}} \rightarrow \bar{\mathbf{D}}$. The question of whether or not g_2 might imply the intuitive causal inference for different VAR modeling parameters was not explored in this work.

Another approach to the practical issues discussed above is to sample smaller contiguous subsets of $\mathbf{D}$, $\mathbf{B}$, and $\mathbf{B}_r$. Let $\hat{\mathbf{D}}^L = \{\{D_{st}(t')\}\}$, $\hat{\mathbf{B}}^L = \{\{B_z(t')\}\}$, and $\hat{\mathbf{B}}_r^L = \{\{B_z'(t')\}\}$ with $t' = t_0', t_1', t_2', \ldots, L$ be an ordered subset of $\mathbf{D}$, $\mathbf{B}$, and $\mathbf{B}_r$. An ECA summary vector could be constructed for each sampled pair $(\hat{\mathbf{B}}^L, \hat{\mathbf{D}}^L)$ and $(\hat{\mathbf{B}}_r^L, \hat{\mathbf{D}}^L)$. A set of n sampled pairs, each with a different t_0', would produce a set of n values for each time series causality tool. These sets of values can then be used to develop a mean ECA summary vector to draw causal inferences.

Let $L = 500$ and $n = 10^4$. The starting points for each time series are sampled from a uniform distribution over $[0, N - L]$. The leaning calculation uses the $(1/4)$-width tolerance domains and the l-standard assignment where l is set as $l = 1, \ldots, l_a$, where l_a is the lag for which

---

[15]Consider the leaning calculated using the 1-standard cause-effect assignment for a given time series pair $(\mathbf{A}, \mathbf{B})$. If the time intervals of the time steps of $\mathbf{A}$ and $\mathbf{B}$ are not uniform, an analyst must be certain to understand the leaning is comparing time steps, not physical time intervals. Thus, $\{C, E\} = \{a_{t-1}, b_t\}$ would not be an assumption of, e.g., an hour in the past of $\mathbf{A}$ drives the present of $\mathbf{B}$ as it would be if $\Delta t = t_n - t_{n-1} = 1$ hour $\forall t_n$ in $\mathbf{A}$ and $\mathbf{B}$.

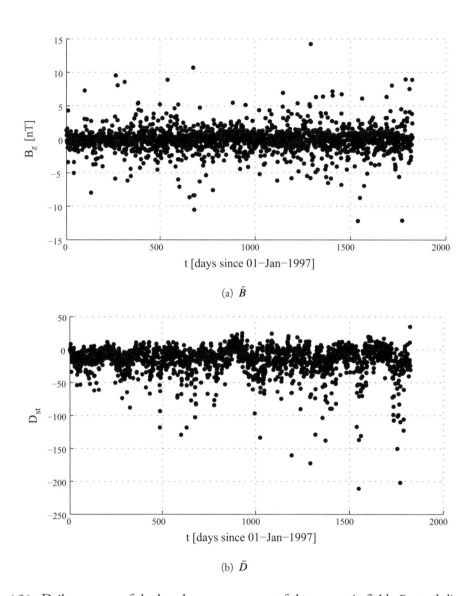

(a) $\bar{B}$

(b) $\bar{D}$

Figure 4.36: Daily averages of the hourly measurements of the magnetic field, B_z, and disturbance storm time index, D_{st} from the beginning of January 1, 1997, until the end of December 31, 2001, taken from the NASA OMNI data set.

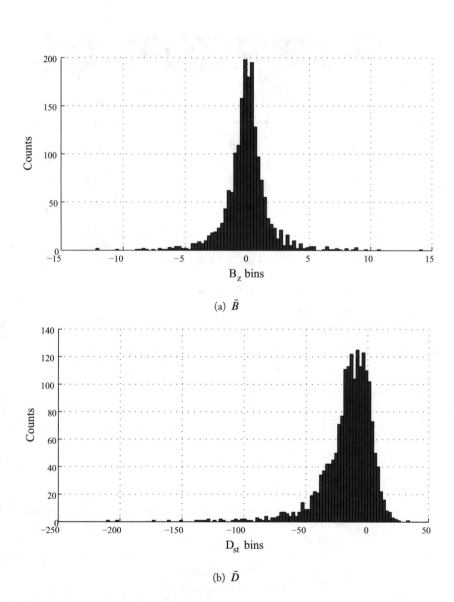

(a) $\bar{B}$

(b) $\bar{D}$

Figure 4.37: Histograms of the data shown in Figure 4.36.

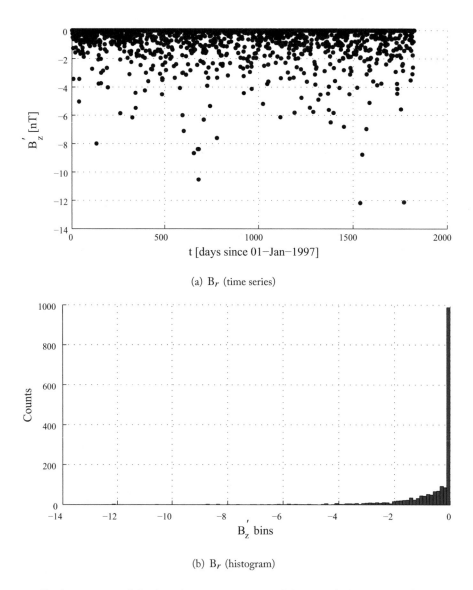

(a) B_r (time series)

(b) B_r (histogram)

Figure 4.38: Daily averages of the hourly measurements of the rectified magnetic field, B_s from the beginning of January 1, 1997, until the end of December 31, 2001, and the 100-bin histogram of that data.

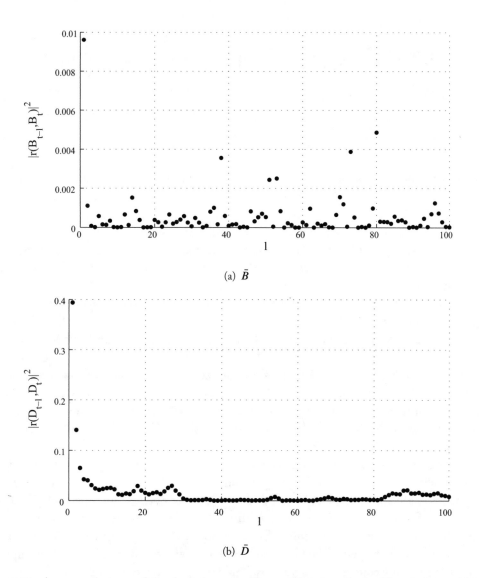

(a) $\bar{B}$

(b) $\bar{D}$

Figure 4.39: Autocorrelations of the data shown in Figure 4.36 given lags of $l = 1, 2, \ldots, 100$. The autocorrelations are $|r(b_{t-l}, b_t)|^2$ where $r(\cdot)$ is the Pearson correlation coefficient between the lagged l series $\{b_{t-l}\}$ and the time series $\{b_t\}$.

Table 4.5: Comparisons of the ECA summaries for each of the time series pairs shown in Figures 4.36 and 4.38 with the intuitive inferences and the "majority-vote" inference, i.e., the causal inference implied by the majority of the times series causality tools used during the exploratory causal analysis

	ECA summary	Majority-vote inference	Intuitive inference
$(\overline{B}, \overline{D})$	undefined	$\overline{B} \to \overline{D}$	$\overline{B} \to \overline{D}$
$(\overline{B}_r, \overline{D})$	$\overline{B}_r \to \overline{D}$	$\overline{B}_r \to \overline{D}$	$\overline{B}_r \to \overline{D}$

the lagged autocorrelation of $\hat{\mathbf{D}}^L$ is minimum in a set of autocorrelations calculated with lags from 1 to $L/2$. The same l used in the leaning calculation will be used in the lagged cross-correlation difference calculations. The PAI correlation differences are calculated with $E = 50$ and $\tau = 1$. Table 4.11 shows the mean value across the n calculated values of each time series causality tools, and Table 4.7 shows a comparison of the implied causal inferences of these mean values and the intuitive inferences. Table 4.9 shows the 90% confidence intervals across the n calculated values of each time series causality tools, and Table 4.8 shows a comparison of the implied causal inferences of these confidence intervals and the intuitive inferences. The 90% confidence interval is defined as $[p_5, p_{95}]$ where p_i is the ith percentile of the data; i.e., the lower bound of the 90% confidence interval is the value that is above 5% of the data and the upper bound of the 90% confidence interval is the value that is above 95% of the data. A bootstrapping [218] procedure can be set up with the sample of n values for each of the time series causality tool calculations, whereby 10^5 means are calculated from new sets (of the same size as the original set) of values that have been sampled (with replacement) from the original set. The 90% confidence interval for the mean of these bootstrapped samples, and the comparisons of the implied causal inference with intuition, are shown in Tables 4.12 and 4.10.

This sampling procedure resulted in five sets of 3,025 values, one set for each of the time series causality tool calculations. Only approximately 3×10^3 of the approximately 10^4 sampled times series subsets $\hat{\mathbf{D}}^L$, $\hat{\mathbf{B}}^L$, and $\hat{\mathbf{B}}_r^L$ (with $L = 500$) contained no missing data. These were the only subsets for which the time series causality tools were calculated. The means of these sets, shown in Table 4.11, lead to undefined ECA summaries but majority-vote inferences that agree with intuition for both pairs. The 90% confidence intervals of these sets, however, imply neither the intuitive nor the counter-intuitive causal inference, as shown in Table 4.9. The bootstrapped 90% confidence intervals, shown in Table 4.12, imply ECA summary vectors that agree with Table 4.11. In both cases, g_1 (transfer entropy) and g_3 (PAI) both imply the counter-intuitive causal inference, while every other element of the ECA summary vector implies the intuitive causal inference.

The exploratory causal analysis of this data, like any exploratory analysis [4], is dependent on the framework within which the causal inferences are drawn. The inferences of Table 4.6

Table 4.6: The values of each of the five time series causality tools used in the exploratory causal analysis for each of the time series pairs shown in Figures 4.36 and 4.38 with the mean leaning across all the lags labeled "L," the mean lagged cross-correlation difference labeled "LCC," Granger causality log-likelihood statistics difference (i.e., $F_{X \to Y} - F_{Y \to X}$, calculated by the MVGC toolbox) labeled "GC," the transfer entropy difference (i.e., $T_{X \to Y} - T_{Y \to X}$, calculated with the JIDT) labeled "TE," and the PAI correlation difference labeled "PAI." An entry of "n/a" in the GC column indicates the MVGC toolbox failed to fit a VAR model to the data with the maximum request model parameters and/or within the maximum allotted computation time.

	TE	GC	PAI	L	LCC	$\vec{g}$	ECA summary
$(\overline{B}, \overline{D})$	3.6×10^{-3}	-4.6×10^{-3}	-1.2×10^{-1}	5.0×10^{-3}	-1.2×10^{-2}	$(0, 1, 0, 0, 0)$	undefined
$(\overline{B}_r, \overline{D})$	2.7×10^{-2}	4.6×10^{-3}	-1.3×10^{-1}	7.6×10^{-3}	-2.5×10^{-2}	$(0, 0, 0, 0, 0)$	$(\overline{B}_r, \to \overline{D})$

Table 4.7: Comparisons of the ECA summaries for each of the time series pairs shown in Table 4.11 with the intuitive inferences and the "majority-vote" inference, i.e., the causal inference implied by the majority of the times series causality tools used during the exploratory causal analysis

	ECA summary	Majority-vote inference	Intuitive inference
$(\hat{B}^L, \hat{D}^L)$	undefined	$\hat{B}^L \to \hat{D}^L$	$\hat{B}^L \to \hat{D}^L$
$(\hat{B}_r^L, \hat{D}^L)$	undefined	$\hat{B}_r^L \to \hat{D}^L$	$\hat{B}_r^L \to \hat{D}^L$

Table 4.8: Comparisons of the ECA summaries for each of the time series pairs shown in Table 4.9

	ECA summary	Majority-vote inference	Intuitive inference
$(\hat{B}^L, \hat{D}^L)$	undefined	undefined	$\hat{B}^L \to \hat{D}^L$
$(\hat{B}_r^L, \hat{D}^L)$	undefined	undefined	$\hat{B}_r^L \to \hat{D}^L$

Table 4.9: The 90% confidence intervals of the average values shown in Table 4.11

	TE	GC	PAI	L
$(\hat{B}^L, \hat{D}^L)$	$[-3.0, -2.9] \times 10^{-2}$	$[1.9, 2.0] \times 10^{-1}$	$[1.2, 1.2] \times 10^{-1}$	$[1.2, 1.2] \times 10^{-1}$
$(\hat{B}_r^L, \hat{D}^L)$	$[-5.5, -5.2] \times 10^{-3}$	$[2.2, 2.2] \times 10^{-1}$	$[9.8, 9.9] \times 10^{-2}$	$[1.1, 1.1] \times 10^{-1}$

	LCC	$\vec{g}$	ECA summary
$(\hat{B}^L, \hat{D}^L)$	$[-1.6, -1.6] \times 10^{-1}$	$(1, 0, 1, 0, 0)$	undefined
$(\hat{B}_r^L, \hat{D}^L)$	$[-1.9, -1.9] \times 10^{-1}$	$(1, 0, 1, 0, 0)$	undefined

Table 4.10: Comparisons of the ECA summaries for each of the time series pairs shown in Table 4.12

	ECA summary	Majority-vote inference	Intuitive inference
$(\hat{B}^L, \hat{D}^L)$	undefined	$\hat{B}^L \to \hat{D}^L$	$\hat{B}^L \to \hat{D}^L$
$(\hat{B}_r^L, \hat{D}^L)$	undefined	$\hat{B}_r \to \hat{D}^L$	$\hat{B}_r^L \to \hat{D}^L$

were drawn from an averaged times series subset of the available data. The averaging procedure implies the majority-inference of $\bar{B} \to \bar{D}$ or the ECA summary of $\bar{B}_r \to \bar{D}$ are statements of daily averages only during the time period of 1997 to 2001. These are not statements about the hourly data during the available time period of 1963 to 2012. It may be that the five years used during the averaging procedure are not representative of other five-year subsets of the 50 years of

Table 4.11: The average values of each of the five time series causality tools used in this exploratory causal analysis for each of the time series pairs $(\hat{\mathbf{B}}^L, \hat{\mathbf{D}}^L)$ and $(\hat{\mathbf{B}}_r^L, \hat{\mathbf{D}}^L)$ with the mean leaning across all the lags labeled "L," the mean lagged cross-correlation difference labeled "LCC," Granger causality log-likelihood statistics difference (i.e., $F_{X\to Y} - F_{Y\to X}$, calculated by the MVGC toolbox) labeled "GC," the transfer entropy difference (i.e., $T_{X\to Y} - T_{Y\to x}$, calculated with the JIDT) labeled "TE," and the PAI correlation difference labeled "PAI." An entry of "n/a" in the GC column indicates the MVGC toolbox failed to fit a VAR model to the data with the maximum request model parameters and/or within the maximum allotted computation time.

	TE	GC	PAI	L	LCC	$\vec{g}$	ECA summary
$(\hat{B}^L, \hat{D}^L)$	-2.9×10^{-2}	1.9×10^{-1}	1.2×10^{-1}	1.2×10^{-1}	1.2×10^{-1}	$(1, 0, 1, 0, 0)$	undefined
$(\hat{B}_r^L, \hat{D}^L)$	-5.3×10^{-3}	2.2×10^{-1}	9.8×10^{-2}	1.1×10^{-1}	1.1×10^{-1}	$(1, 0, 1, 0, 0)$	undefined

Table 4.12: The 90% confidence intervals for the set of 10^5 bootstrap calculations of the means shown in Table 4.11

	TE	GC	PAI	L
$(\hat{\mathbf{B}}^L, \hat{\mathbf{D}}^L)$	$[-9.7, -2.9] \times 10^{-2}$	$[7.8 \times 10^{-2}, 3.5 \times 10^{-1}]$	$[4.1 \times 10^{-2}, 2.1 \times 10^{-1}]$	$[-8.3 \times 10^{-3}, 3.4 \times 10^{-1}]$
$(\hat{\mathbf{B}}_r^L, \hat{\mathbf{D}}^L)$	$[-6.5, -5.5] \times 10^{-2}$	$[6.8 \times 10^{-2}, 4.2 \times 10^{-1}]$	$[1.8 \times 10^{-2}, 1.8 \times 10^{-1}]$	$[-9.2 \times 10^{-3}, 3.6 \times 10^{-1}]$

	LCC	$\vec{g}$	ECA summary	
$(\hat{\mathbf{B}}^L, \hat{\mathbf{D}}^L)$	$[-2.6 \times 10^{-1}, -3.9 \times 10^{-2}]$	$(2,0,1,2,0)$	undefined	
$(\hat{\mathbf{B}}_r^L, \hat{\mathbf{D}}^L)$	$[-2.9 \times 10^{-1}, -7.8 \times 10^{-2}]$	$(2,0,1,2,0)$	undefined	

available data. Trends in the hourly time series may not be present in the daily time series. For example, g_1 (transfer entropy) and g_3 (PAI) imply the intuitive causal inference for the daily time series (Table 4.6) and the counter-intuitive causal inference for the sampled hourly times series (Table 4.11). However, the majority-vote inferences for both the daily and sampled hourly time series pairs agree with intuition. The sampling procedure sampled times series of length $L = 500$, which corresponds to approximately 20 days of hourly data. It follows that the majority-vote inferences of $\hat{\mathbf{B}}_r^L \to \hat{\mathbf{D}}^L$ and $\hat{\mathbf{B}}_r^L \to \hat{\mathbf{D}}^L$ are not statements about the time series pairs $(\mathbf{B}, \mathbf{D})$ and $(\mathbf{B}_r, \mathbf{D})$. The causal inferences drawn from the 20-day sampled times series may be the same causal inference that would be drawn from an exploratory causal analysis of the entire time series pairs $(\mathbf{B}, \mathbf{D})$ and $(\mathbf{B}_r, \mathbf{D})$ but such assumptions would need theoretical (or analytical) support that has not been explored in this example. For example, if the sampling procedure is performed again but with $L = 1000$ (i.e., approximately 40 days of data), then the ECA vector[16] for the pair $(\hat{\mathbf{B}}^L, \hat{\mathbf{D}}^L)$ is $\vec{g} = (1, 0, 1, 0, 0)$, which is the same as the $L = 500$ case. However, the ECA vector for the pair $(\hat{\mathbf{B}}_r^L, \hat{\mathbf{D}}^L)$ is $\vec{g} = (0, 0, 1, 0, 0)$, which is different from the $L = 500$ case. Both ECA vectors for the longer sampled subset time series have majority-vote inferences that agree with intuition, just as was found for the $L = 500$ case. But g_1 (transfer entropy) for $(\hat{\mathbf{B}}_r^L, \hat{\mathbf{D}}^L)$ implies the intuitive inference of $\hat{\mathbf{B}}_r^L \to \hat{\mathbf{D}}^L$ for $L = 1000$, which differs both from the g_1 inference for $(\hat{\mathbf{B}}^L, \hat{\mathbf{D}}^L)$ with $L = 1000$ and from the g_1 inference for either pair with $L = 500$. Changing only the length of the sampled time series subsets can change the causal inference implied by g_1 for $\hat{\mathbf{B}}_r^L \to \hat{\mathbf{D}}^L$.

The missing data points have also affected both the daily averaging procedure and the sampling procedure. The averaging procedure involves the arithmetic mean of 24-hours of data where missing data within that 24-hour period was ignored; i.e., if, e.g., only 20-hours of data are available for a given 24-hour period, then the reported daily mean for that period is the arithmetic mean of only 20 data points rather than the expected 24. This process may or may not have have led to daily averages that are representative of the physical system, i.e., the calculated daily averages may be significantly different from the daily averages that might have been calculated if

---

[16]The ECA vector here is found by using the mean value of each time series causality tool to find the implied causal inference, as was done in Table 4.11.

the data was not missing. Such counter-factual concerns, however, are not (and perhaps cannot) be addressed in this example. The missing data also biases the sampling procedure. The sampling procedure involves the random selection of 500 contiguous data points within the time series $\mathbf{B}$, $\mathbf{B}_r$, and $\mathbf{D}$. If any of the subset time series are missing data, then the exploratory causal analysis is not performed and different subset time series are selected. This is why there are only 3,025 samples of each time series causality tool calculation in the attempted sampling of 10^4 time series subsets.[17] It may be that 500 contiguous data points containing no missing data across all three time series are relatively rare within the data, which may cause the sampling procedure to over-represent certain 20-day time periods in the analysis. This issue was also not addressed in this example.

The majority-vote inferences of both the daily average and sampling procedures used in this example agree with intuition. This analysis, however, is exploratory causal analysis and should not be confused with confirmatory causal analysis. The issues discussed in the previous two paragraphs help illustrate the distinction between the two. The results presented in this section do not confirm any theory that posits $\mathbf{B} \rightarrow \mathbf{D}$. Rather, the results show $\bar{\mathbf{B}} \rightarrow \bar{\mathbf{D}}$ and $\hat{\mathbf{B}}_r^L \rightarrow \hat{\mathbf{D}}^L$ (for $L = 500$) are potential causal structures of the time series pairs $(\bar{\mathbf{B}}, \bar{\mathbf{D}})$ and $(\hat{\mathbf{B}}_r^L, \hat{\mathbf{D}}^L)$ (for $L = 500$), respectively. These results may be useful for the confirmatory causal analysis of $\mathbf{B} \rightarrow \mathbf{D}$ but would require, as stated previously, outside theoretical (or analytical) support.

---

[17]Only 1,366 samples were found in 5,800 attempts when the sampling procedure was run with $L = 1000$.

CHAPTER 5

Conclusions

5.1 ECA RESULTS AND EFFICACY

The causal inferences drawn from every tested time series causality tool agree with intuition for four of the five bivariate synthetic data examples, including every linear example and certain parameters for both nonlinear examples. PAI and transfer entropy lead to counter-intuitive causal inferences for specific parameter values of the coupled logistic map example.[1] PAI and the Granger causality log-likelihood statistic can fail to be defined (and, thus, provide no causal inference) for the RL circuit example (depending on the sampling frequency of the voltage signal).[2] Overall, PAI provides intuitive causal inferences as often as transfer entropy, when it is defined, and is defined as often as the Granger tool. These results indicate that the PAI algorithm introduced in [36] may need to be refined (to handle zero distance nearest neighbors better) and the algorithm parameters (i.e., the embedding dimension, time delay, and number of weights to use in the cross-mapping procedure) may need to be set more carefully during the ECA process. The leaning never fails to be defined for any of the examples and provides the intuitive causal inference for a majority of the tested parameter values for both nonlinear examples. The leaning even provides intuitive causal inferences for examples in which both the transfer entropy and PAI provide counter-intuitive inferences.[3]

The leaning is the only tested time series causality tool that consistently provides the intuitive causal inference for both empirical data examples. The Granger tool and lagged cross-correlation are the only tested tools that imply counter-intuitive causal inferences for the snowfall example.[4] The second empirical example, the OMNI data, shows how the time series sampling procedure can affect the implied causal inferences. The tested time series causality tools did not all provide the same causal inference for either empirical example, which implies the empirical data sets do not have clear, intuitive driving relationships in the sense of the simpler synthetic data examples. The disagreement between the tools implies some complexity in the system dynamics but can also provide additional inferences. For example, the transfer entropy and Granger tool provide conflicting causal inferences for both the snowfall example and most of the sampling procedures used for the OMNI data example. It is known that these two tools are equivalent if the joint distribution of the data is Gaussian [168]. Thus, this disagreement implies the empiri-

---

[1]See Section 4.2.5.
[2]See Section 4.2.3.
[3]See Section 4.2.5.
[4]See Section 4.3.

cal data sets are not jointly Gaussian (i.e., the distributions fail to meet the requirements of the Barnett et al. equivalence proof [168]).

The exploratory causal analysis leads to ECA summaries that agree with intuition for some of the synthetic data examples, e.g., Section 4.2.1, without much effort on the part of the analyst, i.e., the times series causality tools could be applied naively and the results are straightforward to interpret. The exploratory causal analysis of other examples, e.g., Section 4.2.5, was more subtle, requiring careful interpretation of the results, especially given conflicting causal inference implications from different tools and determination of "intuitive" causal inferences from system parameters. The ECA summary agreed with intuition for some examples, and in other examples only the majority-vote inference agreed with intuition. In some examples, e.g., Section 4.2.6, neither agreed with intuition and may have implied counter-intuitive causal inferences. In every example, however, the exploratory causal analysis provided insight into the data that was potentially useful (or, at least, interesting) to an analyst.

This work has posited that exploratory causal analysis should be a part of the exploratory data analysis of time series data sets. Such analysis is particularly useful with data collected from systems for which more traditional physics experiments are not possible, including space weather data sets used for geomagnetic storm forecasting. This work has not explored the detailed exploratory causal analysis that may guide a confirmatory data analysis of such data sets. For example, testing different cause-effect assignments for the leaning calculations of Section 4.3 may help guide the development of structural models (i.e., SEM) that might be used during the confirmatory data analysis of that data (e.g., using SCM).[5] Broadening the scope of the exploratory causal analysis for the example of Section 4.3, e.g., by calculating an ECA summary for every time series pair available in the NASA OMNI data set, may also be fruitful in guiding theoretical or confirmatory causal analysis of the system.

The ECA summary vector, and the exploratory causal analysis approach in general, also addresses the practicing scientist's problem introduced in the preface. The time series causality tools used in every example of this work can be calculated using open source software readily available to a scientist seeking a "plug-and-play" solution to their causal analysis problems.[6] Consider the recent result presented in [192] showing the causal interaction between the Indian Ocean Dipole (IOD) and El Niño. It is shown that the information flow between time series of IOD and El Niño indexes implies a dominant causal structure of "IOD causes El Niño"[7] [192]. The time series causality tools discussed in this work (i.e., specifically the software implementations used in the examples) can be applied to these same time series to draw the same conclusion; i.e., $|\vec{g}|^2 = 0$ if $\mathbf{X}$ is the IOD index and $\mathbf{Y}$ is either El Niño index used in [192]. This ECA summary vector is produced with naive algorithm parameters of $E = 100$, $\tau = 1$, 1/4-width tolerance domains, and l-standard cause-effect assignments with $l = 1, 2, \ldots, 20$ (which are also the lags used for

---

[5]See Section 2 for discussions of SEM and SCM.
[6]See https://github.com/jmmccracken for all of the code used in this work.
[7]The author presents a detailed analysis of this causal structure, but the question of which series might be considered the more dominant driver is the focus of exploratory causal analysis, so this is the only result discussed here.

g_5). All of these parameters are set based on previous experience in generating ECA summary vectors and not from any analysis of the times series themselves. The point is not to imply that ECA summary vectors can replace more in-depth causal analysis of data (see Sections 1.3 and 4 for previous discussions of this point). Rather, the point is that the approach to exploratory causal analysis presented in this work provides tools for an analyst to make causal inferences, and start the process of a more formal causal analysis, without significant effort in developing the necessary tools, such as the derivation of an equation for information flow from transfer entropy using specific assumptions about the systems as was done in [192].

5.2   FUTURE WORK

Several examples of exploratory causal analysis have been shown, some with successful naive ECA summaries and some that required individual time series analysis tools to be applied in different ways. Overall, the theme of using more than one tool to draw exploratory causal inferences proved to be useful both in understanding the potential causal structure of the system and in understanding any failures of particular tools (i.e., disagreements with intuition and/or disagreements with the majority of the other tools). This approach potentially helps prevent errant causal inferences at the cost of (potentially) redundant calculations. This work, however, has not explored these ideas formally. What properties of the system dynamics consistently lead to counter-intuitive causal inferences from one time series causality tool but not another? Are there such properties? These types of research questions may help an analyst better understand the time series causality tools, but may also help guide the system modeling (e.g., if a given Granger causality tool is known to always fail if the time series data is generated by non-linear dynamics, then the failure of such a tool may guide the analyst to eliminate linear approaches during the system modeling effort). This work briefly noted in Section 4.2.4 that a given Granger causality tool (i.e., the MVGC implementation of the Granger log-likelihood statistic) consistently implied the intuitive causal inference for synthetic data sets generated by non-linear dynamics, despite the expectation to not do so (which also happened for the example shown in Section 4.2.5). The exploration of such research questions, however, may require more than computational testing. It may be fruitful to formally explore, e.g., how a given cause-effect assignment in the leaning calculation might appear in the VAR forecast model used by a given Granger causality tool.

The synthetic data example of Section 4.2.5 is the only example in this work of exploratory causal analysis on coupled system dynamics. There is a history of studying such systems in non-linear and chaotic time series analysis (see, e.g., [138]). Such systems, however, do not always have clear physical causal intuitions, so this work did not explore them in detail (i.e., beyond what is done in Section 4.2.5). These systems may provide an intriguing study of the exploratory causal analysis approach. Is it possible for multiple time series analysis tools to imply the same causal inference for such systems? If so, how are the tools that agree related, e.g., formally or computationally? Exploratory causal analysis of well-studied chaotic dynamics may help guide analysts in applying such techniques to, e.g., building forecast models, which is known to be difficult for data generated from non-linear and/or chaotic dynamics [228].

The leaning was introduced in Section 3.5 and used throughout this work. All leaning calculations rely on cause-effect assignment and tolerance domains, which, given empirical data, must be set with some reasonable data analysis. This work used the (1/4)-width tolerances domains extensively, which are straightforward to define for a given data set. The l-standard cause-effect assignments were used exclusively in this work with l set as a function of the autocorrelation lengths of one or both times series. Section 4.2.1 discussed one possible method for setting l algorithmically. There was no discussion, however, of determining a reasonable cause-effect assignment algorithmically in general. The l-standard assignment is a straightforward choice but an analyst may be limiting the usefulness of the leaning calculation by not trying other possible cause-effect assignments. Consider, for example, a system for which one signal $\mathbf{D}$ is a steady impulse representing the times at which a grain of sand is dropped into a pile, i.e., $\mathbf{D} = \{d_t = 1 \ \forall t \in \mathcal{D}, d_t = 0 \ \forall t \notin \mathcal{D}\}$ where $\mathcal{D}$ is the set of times at which the grain of sand is dropped, and the other signal $\mathbf{H}$ is the maximum height of the pile, i.e., $\mathbf{H} = \{h_t = f(d_i) \,|i = 1, 2, \ldots, t\}$ where the maximum height of the pile at time t is a function f of all the previously dropped grains of sand d_i. It is expected that f is sufficiently complex to represent both the height increases due to each additional grain of sand and the occasional height decreases due to sand avalanches when the pile meets certain physical requirements. The intuitive causal inference for this scenario is $\mathbf{D} \to \mathbf{H}$ but the leaning may not be able to imply such an inference using only the l-standard assignment because of the occasional avalanches. Instead, the leaning may require an "autoregressive" cause-effect assignment of $\{C, E\} = \{d_{t-l} \text{ and } h_{t-k}, h_t\}$, i.e., the assumed cause is defined using both the l lagged time steps of the impulse signal $\mathbf{D}$ and the k lagged time steps of the response signal $\mathbf{H}$, to return the intuitive causal inference. Perhaps instead there is some fixed value h_0 of h_t that should be used in the cause-effect assignment somehow, e.g., $\{C, E\} = \{d_{t-l} \text{ and } (h_{t-1} < h_0), h_t\}$. It would be useful to algorithmically determine which cause-effect assignments might be useful for the leaning calculation. This task may be prohibitively difficult,[8] but it may allow the leaning to provide more detailed causal inference, e.g., by identifying which cause-effect assignment leads to a higher leaning than another. Likewise, it may be useful to algorithmically set the embedding dimension and time delay in the PAI correlation difference calculation, e.g., by finding the embedding dimension and time delay that maximum the SSR correlation (i.e., the correlation between a given time series and the estimation of that time series calculated using the shadow manifold of itself). If all the time series causality tool parameters were set algorithmically, directly from the data being analyzed, then the exploratory causal analysis may become significantly automated.

Section 1.3 emphasized that exploratory causal analysis of bi-variate time series data was focused on answering the question "Given $(\mathbf{A}, \mathbf{B})$, is the potential driving relationship $\mathbf{A} \to \mathbf{B}$ or $\mathbf{B} \to \mathbf{A}$, or can no conclusion be drawn with the tools being used?" This focus may be a myopic use

---

[8]The leaning calculation is, as discussed in Section 3.5, essentially a structured counting of features in the time series pairs. So, algorithmically setting a cause-effect assignment for the leaning calculation may be equivalent to pattern finding and matching within and between the two series, and then calculating the leaning for all potential cause-effect assignments.

of the time series tools. The actual values calculated with each tool was only used for comparison to another value and was distilled into a ternary (yes-no-unknown) answer. A fruitful extension of this work may be to explore how the actual values calculated with each of these tools compare to each other and to the causal intuitions drawn from system parameter values in the synthetic data examples. Consider, for example, an instance of Eq. (4.9) with some fixed $x_0 = y_0$, $r_x > r_y$ and $\beta_{yx} > \beta_{xy}$. It was shown in Section 4.2.5 that such an instance of Eq.(4.9) can lead to an ECA summary that agrees with the intuitive causal inference of $\mathbf{X} \to \mathbf{Y}$. If a second instance of Eq. (4.9) is generated with all the same system parameters expect β_{yx}, which is increased by some amount Δ, do the individual values of the times series causality tools also change appropriately by some amount related to Δ? Does, for example, the transfer entropy difference become more positive by some amount related to Δ and/or related to the amount by which the PAI correlation difference becomes more negative? These types of research questions may lead to a deeper understanding of both the time series causality tools themselves and what type of "causality" is actually being investigated by these tools.

Finally, this work has also purposefully ignored the philosophical foundations of the time series causality tools being used. Some of these tools have been discussed in the philosophical literature (e.g., Granger causality and information-theoretic causality tools; see [5]) but others, particularly the leaning, have not. The penchant is derived from probabilistic causality notions and is related to quantities that have been discussed philosophically, including Kleinberg's causal significance, for which the foundational causality issues are discussed at length in [31]. The derivation of the leaning from the penchants and the penchants' relationship to the causal significance may imply that the leaning has an interpretation within Kleinberg's causal framework.

Bibliography

[1] P. Godfrey-Smith. *Theory and Reality: An Introduction to the Philosophy of Science.* Science and Its Conceptual Foundations series. University of Chicago Press, 2009. DOI: 10.7208/chicago/9780226300610.001.0001. 1, 2, 14

[2] G. van Belle. *Statistical Rules of Thumb.* Wiley Series in Probability and Statistics. Wiley, 2011. DOI: 10.1002/9780470377963. 1

[3] R. A. Fisher. *The design of experiments*, volume 12. Oliver and Boyd Edinburgh, 1960. DOI: 10.1136/bmj.1.3923.554-a. 1, 2, 14, 15, 16

[4] J. W Tukey. Exploratory data analysis. 1977. 1, 4, 5, 6, 102

[5] P. Illari and F. Russo. *Causality: Philosophical Theory Meets Scientific Practice.* Oxford University Press, 2014. DOI: 10.5860/choice.190085. 2, 5, 6, 11, 12, 15, 16, 41, 113

[6] C. W. J. Granger. Testing for causality: A personal viewpoint. *Journal of Economic Dynamics and Control*, 2(0):329–352, 1980. DOI: 10.1016/0165-1889(80)90069-X. 2, 8, 11, 23, 24, 41

[7] C. Granger. Time series analysis, cointegration, and applications. *Nobel Lecture*, pages 360–366, 2003. DOI: 10.1257/0002828041464669. 2, 3, 11, 23

[8] R. A. Fisher. Statistical methods for research workers. 1934. DOI: 10.1007/978-1-4612-4380-9_6. 2

[9] G. W. Imbens and D. B. Rubin. *Causal Inference in Statistics, Social, and Biomedical Sciences.* Causal Inference for Statistics, Social, and Biomedical Sciences: An Introduction. Cambridge University Press, 2015. DOI: 10.1017/cbo9781139025751. 2, 5, 12, 14, 15, 16

[10] S. L. Morgan and C. Winship. *Counterfactuals and Causal Inference.* Analytical Methods for Social Research. Cambridge University Press, 2014. DOI: 10.1017/cbo9780511804564. 2, 14, 15, 16, 18

[11] P. Illari, F. Russo, and J. Williamson. *Causality in the Sciences.* OUP Oxford, 2011. DOI: 10.1093/acprof:oso/9780199574131.001.0001. 2, 11, 12, 17

[12] D. Bohm. *Causality and Chance in Modern Physics.* Pennsylvania paperbacks. University of Pennsylvania Press, Incorporated, 1971. DOI: 10.1063/1.3060163. 2, 6, 11, 12, 13, 14

[13] M. Bunge. *Causality and Modern Science.* Courier Corporation, 1979. 2, 6, 11, 12, 13, 14

[14] J. H. King and N. E. Papitashvili. Solar wind spatial scales in and comparisons of hourly wind and ace plasma and magnetic field data. *Journal of Geophysical Research: Space Physics (1978–2012),* 110(A2), 2005. DOI: 10.1029/2004ja010649. 2, 93

[15] J. Pearl. *Causality: Models, Reasoning, and Inference.* Cambridge University Press, 2000. DOI: 10.1017/cbo9780511803161. 2, 3, 4, 5, 6, 8, 11, 12, 14, 15, 17, 18, 19, 39

[16] P. W. Holland. Statistics and causal inference. *Journal of the American statistical Association,* 81(396):945–960, 1986. DOI: 10.1080/01621459.1986.10478354. 2, 3, 4, 8, 11, 12, 14, 15, 16, 17

[17] B. Chen and J. Pearl. Regression and causation: A critical examination of six econometrics textbooks. *Real-World Economics Review, Issue,* (65):2–20, 2013. 3

[18] K. A. Bollen and J. Pearl. Eight myths about causality and structural equation models. In *Handbook of Causal Analysis for Social Research,* pages 301–328. Springer, 2013. DOI: 10.1007/978-94-007-6094-3_15. 3, 17, 18

[19] D. Rogosa. A critique of cross-lagged correlation. *Psychological Bulletin,* 88(2):245, 1980. DOI: 10.1037//0033-2909.88.2.245. 3, 37, 38

[20] B. Russell. On the notion of cause. In *Proceedings of the Aristotelian society,* pages 1–26. JSTOR, 1912. DOI: 10.1093/aristotelian/13.1.1. 6, 12

[21] B. Russell. *Human knowledge: Its Scope and its Limits.* New York: Simon & Schuster, 1948. DOI: 10.4324/9780203875353. 12

[22] W. Salmon. *Scientific Explanation and the Causal Structure of the World.* Princeton University Press, 1984.

[23] H. Reichenbach. The philosophy of space and time. *Lorentz and Poincaré Invariance: 100 Years of Relativity,* 8:218, 2001. 6

[24] C. W. J. Granger. Economic processes involving feedback. *Information and Control,* 6(1):28–48, 1963. DOI: 10.1016/s0019-9958(63)90092-5. 7, 21, 23, 24, 26

[25] C. W. J. Granger. Investigating causal relations by econometric models and cross-spectral methods. *Econometrica: Journal of the Econometric Society,* pages 424–438, 1969. DOI: 10.2307/1912791. 7, 23, 24

[26] D. A. Pierce and L. D. Haugh. Causality in temporal systems: Characterization and a survey. *Journal of Econometrics,* 5(3):265–293, 1977. DOI: 10.1016/0304-4076(77)90039-2. 7

[27] G. W. Schwert. Tests of causality: The message in the innovations. *Carnegie-Rochester Conference Series on Public Policy*, 10(0):55–96, 1979. 7, 24

[28] R. L. Jacobs, E. E. Leamer, and M. P. Ward. Difficulties with testing for causation*. *Economic Inquiry*, 17(3):401–413, 1979. DOI: 10.1111/j.1465-7295.1979.tb00538.x. 7

[29] B. Fitelson and C. Hitchcock. Probabilistic measures of causal strength. In Phyllis McKay Illari, Federica Russo, and Jon Williamson, Eds., *Causality in the Sciences*. Oxford University Press, Oxford, 2011. DOI: 10.1093/acprof:oso/9780199574131.001.0001. 8, 38

[30] G. Sugihara, R. May, H. Ye, C. Hsieh, E. Deyle, M. Fogarty, and S. Munch. Detecting causality in complex ecosystems. *Science*, 338(6106):496–500, 2012. DOI: 10.1126/science.1227079. 8, 9, 30, 31, 32, 36, 75

[31] S. Kleinberg. *Causality, Probability, and Time*. Cambridge University Press, 2012. DOI: 10.1017/cbo9781139207799. 8, 12, 19, 38, 113

[32] P. Suppes. *A Probablistic Theory of Causality*. North Holland Publishing Company, 1970. 8, 12, 18, 38, 41

[33] I. J. Good. Causal propensity: A review. *PSA: Proceedings of the Biennial Meeting of the Philosophy of Science Association*, 1984:829–850, 1984. DOI: 10.1086/psaprocbienmeetp.1984.2.192542. 8, 12, 38

[34] T. Schreiber. Measuring information transfer. *Phys. Rev. Lett.*, 85:461–464, 2000. DOI: 10.1103/physrevlett.85.461. 9, 26, 27, 28

[35] S. Kleinberg and B. Mishra. The temporal logic of causal structures. In *Proceedings of the Twenty-Fifth Conference on Uncertainty in Artificial Intelligence*, pages 303–312. AUAI Press, 2009. 9, 19

[36] J. M. McCracken and R. S. Weigel. Convergent cross-mapping and pairwise asymmetric inference. *Phys. Rev. E*, 90:062903, 2014. DOI: 10.1103/physreve.90.062903. 9, 30, 32, 36, 49, 55, 109

[37] A. Falcon. Aristotle on causality. In *Stanford Encyclopedia of Philosophy*. 2008. 11, 12

[38] M. G. Evans. Causality and explanation in the logic of aristotle. *Philosophy and Phenomenological Research*, 19(4):466–485, 1959. DOI: 10.2307/2105115. 11, 12

[39] C. F. Bolduan. The autobiography of science. *American Journal of Public Health and the Nations Health*, 35(10):1090, 1945. DOI: 10.2105/ajph.35.10.1090-a. 11

[40] K. R. Popper, A. F. Petersen, and J. Mejer. *The World of Parmenides: Essays on the Presocratic Enlightenment*. Routledge, 1998. DOI: 10.4324/9781315824482. 11

[41] I. Düring. *Aristotle in the Ancient Biographical Tradition*. Distr.: Almqvist & Wiksell, Stockholm, 1957. 11

[42] A. Plotnitsky. "dark materials to create more worlds:" On causality in classical physics, quantum physics, and nanophysics. *Journal of Computational and Theoretical Nanoscience*, 8(6):983–997, 2011. DOI: 10.1166/jctn.2011.1778. 11

[43] A. Zellner. Philosophy and objectives of econometrics. The interface between econometrics and economic theory. *Journal of Econometrics*, 136(2):331–339, 2007. DOI: 10.1016/j.jeconom.2005.11.001.

[44] A. Zellner. Causality and econometrics. *Carnegie-Rochester Conference Series on Public Policy*, 10:9–54, 1979. DOI: 10.1016/0167-2231(79)90002-2.

[45] A. Zellner. Causality and causal laws in economics. *Journal of Econometrics*, 39(1):7–21, 1988. DOI: 10.1016/0304-4076(88)90038-3. 11

[46] S. M. Shugan. Editorial-causality, unintended consequences and deducing shared causes. *Marketing Science*, 26(6):731–741, 2007. DOI: 10.1287/mksc.1070.0338. 11, 12

[47] Z. Zheng and P. A. Pavlou. Research note-toward a causal interpretation from observational data: A new bayesian networks method for structural models with latent variables. *Information Systems Research*, 21(2):365–391, 2010. DOI: 10.1287/isre.1080.0224. 12

[48] M. J. Druzdzel and H. A. Simon. Causality in Bayesian belief networks. In *Proceedings of the Ninth international conference on Uncertainty in Artificial Intelligence*, pages3–11. Morgan Kaufmann Publishers Inc., 1993. DOI: 10.1016/b978-1-4832-1451-1.50005-6. 11

[49] A. Honore. Causation in the law. In Edward N. Zalta, Ed., *The Stanford Encyclopedia of Philosophy*. Winter edition, 2010. DOI: 10.1145/379437.379789. 12

[50] G. Young, A. W. Kane, K. Nicholson, and D. W. Shuman. *Causality of Psychological Injury*. Springer, 2007. DOI: 10.1007/978-0-387-36445-2. 12

[51] J. Locke. *An Essay Concerning Human Understanding*. Eliz. Holt, 1700. DOI: 10.1093/oseo/instance.00018020. 12

[52] R. Descartes. Discourse on method. *The Philosophical Works of Descartes, tr. by ES Haldane & GRT Ross (Cambridge, 1911-12)*, 1:92, 1830. DOI: 10.1017/cbo9780511805059.004. 12

[53] I. Kant and J. M. D. Meiklejohn. *Critique of Pure Reason*. Bohn's philosophical library. Henry G. Bohn, 1855. DOI: 10.1037/11654-000. 12

[54] D. Hume and L.A. Selby-Bigge. *A Treatise of Human Nature*. Clarendon Press, 1888. DOI: 10.1093/oseo/instance.00032872. 12

[55] J. S. Mill. *A System of Logic, Ratiocinative and Inductive: Being a Connected View of the Principles of Evidence and the Methods of Scientific Investigation*. Harper & Brothers, 1858. DOI: 10.1017/cbo9781139149846. 12

[56] Plato. *The Allegory of the Cave*. P & L Publication, 2010. 12

[57] B. Spinoza. *On the Improvement of the Understanding: The Ethics; Correspondence*. Dover books on philosophy. Dover, 1955. 12

[58] K. Pearson. *Mathematical Contributions to the Theory of Evolution. On Homotyposis in Homologous but Differentiated Organs*. 1903. DOI: 10.1098/rspl.1902.0099. 12, 13

[59] W. C. Salmon. *Causality and Explanation*. Oxford University Press, Oxford, 1998. DOI: 10.1093/0195108647.001.0001. 12

[60] N. Bohr. *Essays 1958-1962 on Atomic Physics and Human Knowledge*. Ox Bow Press, 1963. DOI: 10.1063/1.3051271. 13

[61] N. Bohr. Causality and complementarity. *Philosophy of Science*, 4(3):289–298, 1937. DOI: 10.1086/286465.

[62] P. J. Riggs. *Quantum Causality: Conceptual Issues in the Causal Theory of Quantum Mechanics*, vol. 23. Springer Science & Business Media, 2009. DOI: 10.1007/978-90-481-2403-9. 13

[63] A. Bohm, H. D. Doebner, and P. Kielanowski. *Irreversibility and Causality: Semigroups and Rigged Hilbert Spaces*. Lecture Notes in Physics. Springer Berlin Heidelberg, 2013. DOI: 10.1007/bfb0106772. 13

[64] M. Pawłowski, T. Paterek, D. Kaszlikowski, V. Scarani, A. Winter, and M. Zukowski. Information causality as a physical principle. *Nature*, 461(7267):1101–1104, 2009. DOI: 10.1038/nature08400. 13

[65] A. Kuzmich, A. Dogariu, L. J. Wang, P. W. Milonni, and R. Y. Chiao. Signal velocity, causality, and quantum noise in superluminal light pulse propagation. *Physical Review Letters*, 86(18):3925, 2001. DOI: 10.1103/physrevlett.86.3925. 13

[66] J. A. Smolin and J. Oppenheim. Locking information in black holes. *Physical Review Letters*, 96(8):081302, 2006. DOI: 10.1103/physrevlett.96.081302. 13

[67] G. W. Gibbons and C. A. R. Herdeiro. Supersymmetric rotating black holes and causality violation. *Classical and Quantum Gravity*, 16(11):3619, 1999. DOI: 10.1088/0264-9381/16/11/311. 13

[68] E. C. Zeeman. Causality implies the lorentz group. *Journal of Mathematical Physics*, 5(4):490–493, 1964. DOI: 10.1063/1.1704140. 13

[69] F. J. Tipler. Singularities and causality violation. *Annals of Physics*, 108(1):1–36, 1977. DOI: 10.1016/0003-4916(77)90348-7.

[70] S. Liberati, S. Sonego, and M. Visser. Faster-than-c signals, special relativity, and causality. *Annals of Physics*, 298(1):167–185, 2002. DOI: 10.1006/aphy.2002.6233. 13

[71] G. F. R. Ellis. Physics, complexity and causality. *Nature*, 435(7043):743–743, 2005. DOI: 10.1038/435743a. 13

[72] J. Barbour. *The End of Time: The Next Revolution in Physics*. Oxford University Press, 1999. DOI: 10.5860/choice.38-0352.

[73] L. S. Schulman. *Time's Arrows and Quantum Measurement*. Cambridge monographs on mathematical physics. Cambridge University Press, 1997. DOI: 10.1017/cbo9780511622878.

[74] R. Penrose. *The Emperor's New Mind: Concerning Computers, Minds, and the Laws of Physics*. Popular Science Series. OUP Oxford, 1999. DOI: 10.1119/1.16207. 13, 14

[75] F. Dowker. Causal sets as discrete spacetime. *Contemporary Physics*, 47(1):1–9, 2006. DOI: 10.1080/17445760500356833. 13

[76] L. Bombelli, J. Lee, D. Meyer, and R. D. Sorkin. Space-time as a causal set. *Phys. Rev. Lett.*, 59:521–524, 1987. DOI: 10.1103/physrevlett.59.521. 13

[77] K. S. Thorne. Closed timelike curves. In *General Relativity and Gravitation 1992, Proceedings of the Thirteenth INT Conference on General Relativity and Gravitation, held at Cordoba, Argentina, 28 June–July 4 1992*, page 295. CRC Press, 1993. 13

[78] S. Aaronson and J. Watrous. Closed timelike curves make quantum and classical computing equivalent. In *Proceedings of the Royal Society of London A: Mathematical, Physical and Engineering Sciences*, vol. 465, pages 631–647. The Royal Society, 2009. DOI: 10.1098/rspa.2008.0350. 13

[79] F. Lobo and P. Crawford. Time, closed timelike curves and causality. In *The Nature of Time: Geometry, Physics and Perception*, pages 289–296. Springer, 2003. DOI: 10.1007/978-94-010-0155-7_30. 13

[80] S. M. Korotaev and E. O. Kiktenko. Quantum causality in closed timelike curves. *Physica Scripta*, 90(8):085101, 2015. DOI: 10.1088/0031-8949/90/8/085101. 13

[81] J. G. Cramer and N. Herbert. An inquiry into the possibility of nonlocal quantum communication. *arXiv preprint arXiv:1409.5098*, 2014. 13

[82] E. Mach and B. F. McGuinness. *Principles of the Theory of Heat: Historically and Critically Elucidated*. Vienna Circle Collection. Springer Netherlands, 1986. DOI: 10.1007/978-94-009-4622-4. 13

[83] E. Mach. *The Science of Mechanics*. Cambridge Scholars Publishing, 2009. DOI: 10.1017/cbo9781107338401. 13

[84] P. A. White. Ideas about causation in philosophy and psychology. *Psychological Bulletin*, 108(1):3, 1990. DOI: 10.1037/0033-2909.108.1.3. 13

[85] K. Sawa. Predictive behavior and causal learning in animals and humans1. *Japanese Psychological Research*, 51(3):222–233, 2009. DOI: 10.1111/j.1468-5884.2009.00396.x. 13

[86] A. H. Taylor, R. Miller, and R. D. Gray. New caledonian crows reason about hidden causal agents. *Proceedings of the National Academy of Sciences*, 109(40):16389–16391, 2012. DOI: 10.1073/pnas.1208724109. 13

[87] P. A. White. Causal processing: origins and development. *Psychological Bulletin*, 104(1):36, 1988. DOI: 10.1037//0033-2909.104.1.36. 13

[88] P. A. White. A theory of causal processing. *British Journal of Psychology*, 80(4):431–454, 1989. DOI: 10.1111/j.2044-8295.1989.tb02334.x.

[89] T. R. Shultz. Causal reasoning in the social and nonsocial realms. *Canadian Journal of Behavioural Science/Revue Canadienne des Sciences du Comportement*, 14(4):307, 1982. DOI: 10.1037/h0081266. 13

[90] P. W. Cheng. From covariation to causation: a causal power theory. *Psychological Review*, 104(2):367, 1997. DOI: 10.1037/0033-295x.104.2.367. 13

[91] A. M. Leslie and S. Keeble. Do six-month-old infants perceive causality? *Cognition*, 25(3):265–288, 1987. DOI: 10.1016/S0010-0277(87)80006-9. 13

[92] L. M. Oakes and L. B. Cohen. Infant perception of a causal event. *Cognitive Development*, 5(2):193–207, 1990. DOI: 10.1016/0885-2014(90)90026-p.

[93] A. Michotte, T. R. Miles, and E. Miles. The perception of causality. *British Journal for the Philosophy of Science*, 15(59):254–259, 1964. 13

[94] S. Golin, P. D. Sweeney, and D. E. Shaeffer. The causality of causal attributions in depression: A cross-lagged panel correlational analysis. *Journal of Abnormal Psychology*, 90(1):14, 1981. DOI: 10.1037/0021-843x.90.1.14. 14

[95] W. Yan and E. L. Gaier. Causal attributions for college success and failure an asian-american comparison. *Journal of Cross-Cultural Psychology*, 25(1):146–158, 1994. DOI: 10.1177/0022022194251009. 14

[96] S. P. Nguyen and K. S. Rosengren. Causal reasoning about illness: A comparison between european-and vietnamese-american children. *Journal of Cognition and Culture*, 4(1):51–78, 2004. DOI: 10.1163/156853704323074750. 14

[97] S. A. Sloman and D. Lagnado. Causality in thought. *Annual Review of Psychology*, 66(1):223–247, 2015. PMID: 25061673. DOI: 10.1146/annurev-psych-010814-015135. 14

[98] D. B. Rubin. Estimating causal effects of treatments in randomized and nonrandomized studies. *Journal of Educational Psychology*, 66(5):688, 1974. DOI: 10.1037/h0037350. 16

[99] D. B. Rubin. Statistics and causal inference: Comment: Which ifs have causal answers. *Journal of the American Statistical Association*, 81(396):961–962, 1986. DOI: 10.2307/2289065. 16

[100] J. Neyman and K. Iwaszkiewicz. Statistical problems in agricultural experimentation. *Supplement to the Journal of the Royal Statistical Society*, pages 107–180, 1935. DOI: 10.2307/2983637. 16

[101] A. P. Dawid. Causal inference without counterfactuals. *Journal of the American Statistical Association*, 95(450):407–424, 2000. DOI: 10.1080/01621459.2000.10474210. 16, 17

[102] J. Pearl. Causal inference without counterfactuals: Comment. *Journal of the American Statistical Association*, pages 428–431, 2000. DOI: 10.2307/2669380. 16

[103] C. Granger. Comment. *Journal of the American Statistical Association*, 81(396):967–968, 1986. DOI: 10.1080/01621459.1986.10478358. 16

[104] A. P. Dawid. Influence diagrams for causal modelling and inference. *International Statistical Review*, 70(2):161–189, 2002. DOI: 10.1111/j.1751-5823.2002.tb00354.x. 16, 18

[105] A. P. Dawid. Counterfactuals, Hypotheticals and Potential Responses: A Philosophical Examination of Statistical Causality. In Federica Russo and Jon Williamson, Eds., *Causality and Probability in the Sciences*, pages 503–532. 2007. 16

[106] A. P. Dawid. Beware of the dag! *NIPS Causality: Objectives and Assessment*, 6:59–86, 2010. 16

[107] S. Geneletti and A. P. Dawid. Defining and identifying the effect of treatment on the treated. In Phyllis McKay Illari, Federica Russo, and Jon Williamson, Eds., *Causality in the Sciences*. Oxford University Press, Oxford, 2011. DOI: 10.1093/acprof:oso/9780199574131.001.0001. 17

[108] E. Arjas and M. Eerola. On predictive causality in longitudinal studies. *Journal of Statistical Planning and Inference*, 34(3):361–386, 1993. DOI: 10.1016/0378-3758(93)90146-w. 17

[109] E. Arjas and J. Parner. Causal reasoning from longitudinal data*. *Scandinavian Journal of Statistics*, 31(2):171–187, 2004. DOI: 10.1111/j.1467-9469.2004.02-134.x. 17

[110] M. Eerola. *Probabilistic Causality in Longitudinal Studies*. Lecture Notes in Statistics. Springer, New York, 2012. DOI: 10.1007/978-1-4612-2684-0. 17, 18

[111] T. G. Ditterrich. Machine learning research: four current direction. *Artificial Intelligence Magazine*, 4:97–136, 1997. 17

[112] I. Guyon, C. Aliferis, G. Cooper, A. Elisseeff J.-P. Pellet, P. Spirtes, and A. Statnikov. Causality Workbench. In Phyllis McKay Illari, Federica Russo, and Jon Williamson, Eds., *Causality in the Sciences*. OUP Oxford, 2011. DOI: 10.1093/acprof:oso/9780199574131.001.0001. 17

[113] J. C. Anderson and D. W. Gerbing. Structural equation modeling in practice: A review and recommended two-step approach. *Psychological Bulletin*, 103(3):411, 1988. DOI: 10.1037//0033-2909.103.3.411. 17

[114] J. Pearl. The causal foundations of structural equation modeling. Technical report, DTIC Document, 2012. 18

[115] K. A. Bollen. *Structural Equations with Latent Variables*. Wiley Series in Probability and Statistics. Wiley, 2014. DOI: 10.1002/9781118619179. 17

[116] N. Hall. Structural equations and causation. *Philosophical Studies*, 132(1):109–136, 2007. DOI: 10.1007/s11098-006-9057-9. 18, 19

[117] C. Hitchcock. Structural equations and causation: Six counterexamples. *Philosophical Studies*, 144(3):391–401, 2009. DOI: 10.1007/s11098-008-9216-2. 18

[118] B. J. Biddle and M. M. Marlin. Causality, confirmation, credulity, and structural equation modeling. *Child Development*, 58(1):4–17, 1987. DOI: 10.2307/1130287. 18

[119] N. Cliff. Some cautions concerning the application of causal modeling methods. *Multivariate Behavioral Research*, 18(1):115–126, 1983. DOI: 10.1207/s15327906mbr1801_7. 18

[120] S. Greenland and B. Brumback. An overview of relations among causal modeling methods. *International Journal of Epidemiology*, 31(5):1030–1037, 2002. DOI: 10.1093/ije/31.5.1030. 18

[121] S. L. Lauritzen. *Graphical Models*. Clarendon Press, 1996. 18

[122] S. Wright. The method of path coefficients. *The Annals of Mathematical Statistics*, 5(3):161–215, 1934. DOI: 10.1214/aoms/1177732676. 18

[123] G. U. Yule. Notes on the theory of association of attributes in statistics. *Biometrika*, pages 121–134, 1903. DOI: 10.1093/biomet/2.2.121. 18

[124] P. Spirtes, C. N. Glymour, and R. Scheines. *Causation, Prediction, and Search*. Adaptive computation and machine learning. MIT Press, 2000. DOI: 10.1007/978-1-4612-2748-9. 18

[125] C. Hitchcock. Probabilistic causation. In Edward N. Zalta, Ed., *The Stanford Encyclopedia of Philosophy*. Winter edition, 2012. DOI: 10.1145/379437.379789. 18

[126] J. Pearl et al. Causal inference in statistics: An overview. *Statistics Surveys*, 3:96–146, 2009. DOI: 10.1214/09-ss057. 18, 19

[127] J. Woodward. *Making Things Happen: A Theory of Causal Explanation*. Oxford Studies in the Philosophy of Science. Oxford University Press, 2003. DOI: 10.1093/0195155270.001.0001. 18

[128] K. J. Rothman. Causes. *American Journal of Epidemiology*, 104(6):587–592, 1976. 18

[129] C. Meek and C. Glymour. Conditioning and intervening. *The British Journal for the Philosophy of Science*, 45(4):1001–1021, 1994. DOI: 10.1093/bjps/45.4.1001. 19

[130] S. Greenland. Relation of probability of causation to relative risk and doubling dose: a methodologic error that has become a social problem. *American Journal of Public Health*, 89(8):1166–1169, 1999. DOI: 10.2105/ajph.89.8.1166.

[131] S. R. Cole and M. A. Hernán. Fallibility in estimating direct effects. *International Journal of Epidemiology*, 31(1):163–165, 2002. DOI: 10.1093/ije/31.1.163.

[132] O. A. Arah. The role of causal reasoning in understanding simpson's paradox, lord's paradox, and the suppression effect: covariate selection in the analysis of observational studies. *Emerging Themes in Epidemiology*, 5(1):1–5, 2008. DOI: 10.1186/1742-7622-5-5.

[133] I. Shrier. Propensity scores. *Statistics in Medicine*, 28(8):1317–1318, 2009. DOI: 10.1002/sim.3554.

[134] J. Pearl. Letter to the editor: Remarks on the method of propensity score. *Department of Statistics, UCLA*, 2009. DOI: 10.1002/sim.3521. 19

[135] C. Glymour, D. Danks, B. Glymour, F. Eberhardt, J. Ramsey, R. Scheines, P. Spirtes, C. M. Teng, and J. Zhang. Actual causation: a stone soup essay. *Synthese*, 175(2):169–192, 2010. 19

[136] P. Menzies. Causal models, token causation, and processes. *Philosophy of Science*, 71(5):820–832, 2004. DOI: 10.1086/425057. 19

[137] S. Kleinberg. A logic for causal inference in time series with discrete and continuous variables. In *IJCAI Proceedings-International Joint Conference on Artificial Intelligence*, vol. 22, page 943, 2011. DOI: 10.5591/978-1-57735-516-8/IJCAI11-163. 19

[138] H. Kantz and T. Schreiber. *Nonlinear Time Series Analysis*. Cambridge nonlinear science series. Cambridge University Press, 2004. DOI: 10.1017/cbo9780511755798. 19, 21, 22, 111

[139] G. E. P. Box, G. M. Jenkins, and G. C. Reinsel. *Time Series Analysis: Forecasting and Control*. Wiley Series in Probability and Statistics. Wiley, 2013. DOI: 10.1002/9781118619193. 19, 22, 37, 47

[140] J. Cui, L. Xu, S. L. Bressler, M. Ding, and H. Liang. Bsmart: a matlab/c toolbox for analysis of multichannel neural time series. *Neural Networks*, 21(8):1094–1104, 2008. DOI: 10.1016/j.neunet.2008.05.007. 21

[141] Q. Luo, W. Lu, W. Cheng, P. A. Valdes-Sosa, X. Wen, M. Ding, and J. Feng. Spatiotemporal granger causality: A new framework. *NeuroImage*, 79:241–263, 2013. DOI: 10.1016/j.neuroimage.2013.04.091.

[142] M. G. Tana, R. Sclocco, and A. M. Bianchi. Gmac: A matlab toolbox for spectral granger causality analysis of fmri data. *Computers in Biology and Medicine*, 42(10):943–956, 2012. DOI: 10.1016/j.compbiomed.2012.07.003.

[143] L. Barnett and A. K. Seth. The mvgc multivariate granger causality toolbox: a new approach to granger-causal inference. *Journal of Neuroscience Methods*, 223:50–68, 2014. DOI: 10.1016/j.jneumeth.2013.10.018. 25

[144] Z. Zang, C. Yan, Z. Dong, J. Huang, and Y. Zang. Granger causality analysis implementation on matlab: A graphic user interface toolkit for fmri data processing. *Journal of Neuroscience Methods*, 203(2):418–426, 2012. DOI: 10.1016/j.jneumeth.2011.10.006.

[145] A. K. Seth. A {MATLAB} toolbox for granger causal connectivity analysis. *Journal of Neuroscience Methods*, 186(2):262–273, 2010. DOI: 10.1016/j.jneumeth.2009.11.020. 21

[146] M. J. Kaminski and K. J. Blinowska. A new method of the description of the information flow in the brain structures. *Biological Cybernetics*, 65(3):203–210, 1991. DOI: 10.1007/bf00198091. 21

[147] L. A. Baccala and K. Sameshima. Partial directed coherence: a new concept in neural structure determination. *Biological Cybernetics*, 84(6):463–474, 2001. DOI: 10.1007/pl00007990. 21

[148] K. Hlavackova-Schindler, M. Palus, M. Vejmelka, and J. Bhattacharya. Causality detection based on information-theoretic approaches in time series analysis. *Physics Reports*, 441(1):1–46, 2007. DOI: 10.1016/j.physrep.2006.12.004. 21, 26, 27, 28

[149] A. Kaiser and T. Schreiber. Information transfer in continuous processes. *Physica D: Nonlinear Phenomena*, 166(1-2):43–62, 2002. DOI: 10.1016/s0167-2789(02)00432-3. 21, 22, 26, 27, 28, 29, 30

[150] M. Wibral, R. Vicente, and M. Lindner. Transfer entropy in neuroscience. In M. Wibral, R. Vicente, and J. T. Lizier, Eds., *Directed Information Measures in Neuroscience*, Understanding Complex Systems, pages 3–36. Springer Berlin Heidelberg, 2014. DOI: 10.1007/978-3-642-54474-3. 21

[151] M. Lindner, R. Vicente, V. Priesemann, and M. Wibral. Trentool: A matlab open source toolbox to analyse information flow in time series data with transfer entropy. *BMC Neuroscience*, 12(1):119, 2011. DOI: 10.1186/1471-2202-12-119.

[152] A. Montalto, L. Faes, and D. Marinazzo. Mute: a matlab toolbox to compare established and novel estimators of the multivariate transfer entropy. 2014. DOI: 10.1371/journal.pone.0109462.

[153] J. T. Lizier. Jidt: An information-theoretic toolkit for studying the dynamics of complex systems. *Frontiers in Robotics and AI*, 1(11), 2014. DOI: 10.3389/frobt.2014.00011. 21, 47

[154] N. H. Packard, J. P. Crutchfield, J. D. Farmer, and R. S. Shaw. Geometry from a time series. *Physical Review Letters*, 45(9):712, 1980. DOI: 10.1103/physrevlett.45.712. 22, 30

[155] T. Sauer, J. A. Yorke, and M. Casdagli. Embedology. *Journal of Statistical Physics*, 65(3-4):579–616, 1991. DOI: 10.1007/bf01053745. 22, 30

[156] J. D. Farmer and J. J. Sidorowich. Predicting chaotic time series. *Phys. Rev. Lett.*, 59:845–848, 1987. DOI: 10.1103/physrevlett.59.845. 22, 30

[157] M. Casdagli, S. Eubank, J. D. Farmer, and J. Gibson. State space reconstruction in the presence of noise. *Physica D: Nonlinear Phenomena*, 51(1):52–98, 1991. DOI: 10.1016/0167-2789(91)90222-u. 22

[158] H. Ma and C. Han. Selection of embedding dimension and delay time in phase space reconstruction. *Frontiers of Electrical and Electronic Engineering in China*, 1(1):111–114, 2006. DOI: 10.1007/s11460-005-0023-7. 22, 32

[159] M. Ataei, A. Khaki-Sedigh, B. Lohmann, and C. Lucas. Determination of embedding dimension using multiple time series based on singular value decomposition. In *Proceedings of*

the Fourth International Symposium on Mathematical Modeling. Vienna, Austria, pages 190–196, 2003. 90

[160] M. Small and C. K. Tse. Optimal embedding parameters: a modeling paradigm. *Physica D: Nonlinear Phenomena*, 194(3):283–296, 2004. DOI: 10.1016/j.physd.2004.03.006. 32

[161] M. B. Kennel, R. Brown, and H. D. I. Abarbanel. Determining embedding dimension for phase-space reconstruction using a geometrical construction. *Phys. Rev. A*, 45:3403–3411, 1992. DOI: 10.1103/physreva.45.3403. 22, 32, 90

[162] M. C. Maher and R. D. Hernandez. Causemap: Fast inference of causality from complex time series. *PeerJ*, 3:e824, 2015. DOI: 10.7717/peerj.824. 22

[163] M. El-Gohary and J. McNames. Establishing causality with whitened cross-correlation analysis. *Biomedical Engineering, IEEE Transactions on*, 54(12):2214–2222, 2007. DOI: 10.1109/tbme.2007.906519. 22, 37, 38

[164] C. W. J. Granger. Some recent development in a concept of causality. *Journal of Econometrics*, 39(1-2):199–211, 1988. DOI: 10.1016/0304-4076(88)90045-0. 23

[165] J. Lin. Notes on testing causality, May 2008. 24

[166] M. Ding, Y. Chen, and S. L. Bressle. Granger causality: Basic theory and application to neuroscience. In Jens Timmer Bjorn Schelter, Matthias Winterhalder, Ed., *Handbook of Time Series Analysis*. WILEY-VCH Verlag GmbH & Co, KGaA, Weinheim, 2006. DOI: 10.1002/9783527609970. 24

[167] H. Y. Toda and P. C. B. Phillips. Vector autoregression and causality: a theoretical overview and simulation study. *Econometric Reviews*, 13(2):259–285, 1994. DOI: 10.1080/07474939408800286. 24, 25

[168] L. Barnett, A. B. Barrett, and A. K. Seth. Granger causality and transfer entropy are equivalent for gaussian variables. *Phys. Rev. Lett.*, 103:238701, 2009. DOI: 10.1103/physrevlett.103.238701. 25, 28, 109, 110

[169] C. Berzuini, P. Dawid, and L. Bernardinell. *Causality: Statistical Perspectives and Applications*. Wiley Series in Probability and Statistics. Wiley, 2012. DOI: 10.1002/9781119945710. 26

[170] Z. He and K. Maekawa. On spurious granger causality. *Economics Letters*, 73(3):307–313, 2001. DOI: 10.1016/s0165-1765(01)00498-0. 26

[171] Y. Chen, G. Rangarajan, J. Feng, and M. Ding. Analyzing multiple nonlinear time series with extended granger causality. *Physics Letters A*, 324(1):26–35, 2004. DOI: 10.1016/j.physleta.2004.02.032. 26

[172] N. Ancona, D. Marinazzo, and S. Stramaglia. Radial basis function approach to nonlinear granger causality of time series. *Phys. Rev. E*, 70:056221, 2004. DOI: 10.1103/physreve.70.056221. 26

[173] W. A. Freiwald, P. Valdes, J. Bosch, R. Biscay, J. C. Jimenez, L. M. Rodriguez, V. Rodriguez, A. K. Kreiter, and W. Singer. Testing non-linearity and directedness of interactions between neural groups in the macaque inferotemporal cortex. *Journal of Neuroscience Methods*, 94(1):105–119, 1999. DOI: 10.1016/s0165-0270(99)00129-6. 26

[174] W. Hesse, E. Moller, M. Arnold, and B. Schack. The use of time-variant {EEG} granger causality for inspecting directed interdependencies of neural assemblies. *Journal of Neuroscience Methods*, 124(1):27–44, 2003. DOI: 10.1016/s0165-0270(02)00366-7. 26

[175] M. Kaminski, M. Ding, W. A. Truccolo, and S. L. Bressler. Evaluating causal relations in neural systems: Granger causality, directed transfer function and statistical assessment of significance. *Biological Cybernetics*, 85(2):145–157, 2001. DOI: 10.1007/s004220000235.

[176] L. Barnett and A. K. Seth. Granger causality for state-space models. *Phys. Rev. E*, 91:040101, 2015. DOI: 10.1103/physreve.91.040101. 26, 47

[177] A. Arnold, Y. Liu, and N. Abe. Temporal causal modeling with graphical granger methods. In *Proceedings of the 13th ACM SIGKDD International Conference on Knowledge Discovery and Data Mining*, KDD '07, pages 66–75, New York, NY, USA, 2007. ACM. DOI: 10.1145/1281192.1281203. 26

[178] A. Shojaie and G. Michailidis. Discovering graphical granger causality using the truncating lasso penalty. *Bioinformatics*, 26(18):i517–i523, 2010. DOI: 10.1093/bioinformatics/btq377. 26

[179] A. C. Lozano, N. Abe, Y. Liu, and S. Rosset. Grouped graphical granger modeling for gene expression regulatory networks discovery. *Bioinformatics*, 25(12):i110–i118, 2009. DOI: 10.1093/bioinformatics/btp199. 26

[180] M. Dhamala, G. Rangarajan, and M. Ding. Estimating granger causality from fourier and wavelet transforms of time series data. *Phys. Rev. Lett.*, 100:018701, 2008. DOI: 10.1103/physrevlett.100.018701. 26

[181] C. E. Shannon. A mathematical theory of communication. *Bell System Technical Journal, The*, 27(3):379–423, 1948. DOI: 10.1002/j.1538-7305.1948.tb01338.x. 26

[182] S. Ghosh, K. P. Burnham, N. F. Laubscher, G. E. Dallal, L. Wilkinson, D. F. Morrison, M. W. Loyer, B. Eisenberg, S. Kullback, I. T. Jolliffe, and J. S. Simonoff. Letters to the editor. *The American Statistician*, 41(4):338–341, 1987. 27

[183] S. Kullback and R. A. Leibler. On information and sufficiency. *Ann. Math. Statist.*, 22(1):79–86, 1951. DOI: 10.1214/aoms/1177729694. 27

[184] E. T. Jaynes and G. L. Bretthorst. *Probability Theory: The Logic of Science.* Cambridge University Press, 2003. DOI: 10.1017/cbo9780511790423. 27, 35

[185] P. Amblard and O. J. J. Michel. The relation between granger causality and directed information theory: a review. *Entropy*, 15(1):113, 2012. DOI: 10.3390/e15010113. 29

[186] M. Lungarella, K. Ishiguro, Y. Kuniyoshi, and N. Otsu. Methods for quantifying the causal structure of bivariate time series. *International Journal of Bifurcation and Chaos*, 17(03):903–921, 2007. DOI: 10.1142/s0218127407017628. 29

[187] J. Runge, J. Heitzig, N. Marwan, and J. Kurths. Quantifying causal coupling strength: A lag-specific measure for multivariate time series related to transfer entropy. *Phys. Rev. E*, 86:061121, 2012. DOI: 10.1103/physreve.86.061121. 29, 38

[188] D. Janzing, D. Balduzzi, M. Grosse-Wentrup, and B. Scholkopf. Quantifying causal influences. *The Annals of Statistics*, 41(5):2324–2358, 2013. DOI: 10.1214/13-aos1145. 29

[189] J. M. Nichols. Examining structural dynamics using information flow. *Probabilistic Engineering Mechanics*, 21(4):420–433, 2006. DOI: 10.1016/j.probengmech.2006.02.003. 30

[190] X. S. Liang. Information flow within stochastic dynamical systems. *Phys. Rev. E*, 78:031113, 2008. DOI: 10.1103/physreve.78.031113. 30

[191] X. S. Liang. The liang-kleeman information flow: Theory and applications. *Entropy*, 15(1):327–360, 2013. DOI: 10.3390/e15010327.

[192] X. S. Liang. Unraveling the cause-effect relation between time series. *Phys. Rev. E*, 90:052150, 2014. DOI: 10.1103/physreve.90.052150. 30, 110, 111

[193] E. R. Deyle and G. Sugihara. Generalized theorems for nonlinear state space reconstruction. *PLoS ONE*, 6(3):e18295, 2011. DOI: 10.1371/journal.pone.0018295. 30

[194] Floris Takens. Detecting strange attractors in turbulence. *Dynamical Systems and Turbulence*, 898:366–381, 1981. DOI: 10.1007/bfb0091924. 30

[195] G. Sugihara, B. Grenfell, R. M. May, P. Chesson, H. M. Platt, and M. Williamson. Distinguishing error from chaos in ecological time series and discussion. *Philosophical Transactions of the Royal Society of London. Series B: Biological Sciences*, 330(1257):235–251, 1990. DOI: 10.1098/rstb.1990.0195. 30, 31

[196] G. Sugihara and R. M. May. Nonlinear forecasting as a way of distinguishing chaos from measurement error in time series. *Nature*, 344:734–741, 1990. DOI: 10.1038/344734a0. 30, 31

[197] H. Ye, E. R. Deyle, L. J. Gilarranz, and G. Sugihara. Distinguishing time-delayed causal interactions using convergent cross mapping. *Scientific reports*, 5:14750–14750, 2014. DOI: 10.1038/srep14750. 30

[198] E. R. Deyle, M. Fogarty, C. Hsieh, L. Kaufman, A. D. MacCall, S. B. Munch, C. T. Perretti, H. Ye, and G. Sugihara. Predicting climate effects on pacific sardine. *Proceedings of the National Academy of Sciences*, 110(16):6430–6435, 2013. DOI: 10.1073/pnas.1215506110. 30, 32

[199] X. Wang, S. Piao, P. Ciais, P. Friedlingstein, R. B. Myneni, P. Cox, M. Heimann, J. Miller, S. Peng, T. Wang, et al. A two-fold increase of carbon cycle sensitivity to tropical temperature variations. *Nature*, 2014. DOI: 10.1038/nature12915. 30

[200] J. R. Anastas. *Individual Differences on Multi-Scale Measures of Executive Function*. Ph.D. thesis, University of Connecticut, 2013. 30

[201] A. Bozorgmagham and S. Ross. Dynamical system tools and causality analysis. In *SIAM (Society for Industrial and Applied Mathematics), Student Chapter at Virginia Tech*, 2013. 30

[202] I. Vlachos and D. Kugiumtzis. State space reconstruction from multiple time series. In *Topics on Chaotic Systems: Selected Papers from Chaos 2008 International Conference*, page 378. World Scientific, 2009. DOI: 10.1142/9789814271349_0043. 32

[203] H. Ma, K. Aihara, and L. Chen. Detecting causality from nonlinear dynamics with short-term time series. *Scientific Reports*, 4, 2014. DOI: 10.1038/srep07464. 36

[204] B. Cummins, T. Gedeon, and K. Spendlove. On the efficacy of state space reconstruction methods in determining causality. *SIAM Journal on Applied Dynamical Systems*, 14(1):335–381, 2015. DOI: 10.1137/130946344. 36

[205] J. Park and I. W. Sandberg. Universal approximation using radial-basis-function networks. *Neural Computation*, 3(2):246–257, 1991. DOI: 10.1162/neco.1991.3.2.246. 36

[206] T. Clark, A, H. Ye, F. Isbell, E. R. Deyle, J. Cowles, G. D. Tilman, and G. Sugihara. Spatial convergent cross mapping to detect causal relationships from short time series. *Ecology*, 96(5):1174–1181, 2015. DOI: 10.1890/14-1479.1. 36

[207] L. M. Pecora, T. L. Carroll, and J. F. Heagy. Statistics for mathematical properties of maps between time series embeddings. *Phys. Rev. E*, 52:3420–3439, 1995. DOI: 10.1103/physreve.52.3420. 36

[208] D. A. Kenny. Cross-lagged panel correlation: A test for spuriousness. *Psychological Bulletin*, 82(6):887, 1975. DOI: 10.1037//0033-2909.82.6.887. 37

[209] M. S. Bartlett. Some aspects of the time-correlation problem in regard to tests of signifi-cance. *Journal of the Royal Statistical Society*, pages 536–543, 1935. DOI: 10.2307/2342284. 37

[210] J. Runge, V. Petoukhov, and J. Kurths. Quantifying the strength and delay of climatic interactions: the ambiguities of cross correlation and a novel measure based on graphical models. *Journal of Climate*, 27(2):720–739, 2014. DOI: 10.1175/jcli-d-13-00159.1. 37, 38

[211] G. C. Carter. Coherence and time delay estimation. *Proceedings of the IEEE*, 75(2):236–255, 1987. DOI: 10.1109/proc.1987.13723. 37

[212] R. M. Rozelle and D. T. Campbell. More plausible rival hypotheses in the cross-lagged panel correlation technique. *Psychological Bulletin*, 71(1):74, 1969. DOI: 10.1037/h0026863. 37

[213] A. H. Yee and N. L. Gage. Techniques for estimating the source and direction of causal influence in panel data. *Psychological Bulletin*, 70(2):115, 1968. DOI: 10.1037/h0025927. 37

[214] D. L. Weakliem. A critique of the Bayesian information criterion for model selection. *Sociological Methods & Research*, 27(3):359–397, 1999. DOI: 10.1177/0049124199027003002. 47

[215] B. Hayes. Third base. *American Scientist*, 89(6), 2001. DOI: 10.1511/2001.40.3268. 48

[216] D. Halliday, R. Resnick, and J. Walker. *Fundamentals of Physics*. John Wiley & Sons, 2010. DOI: 10.1063/1.3070817. 64

[217] R. D. Knight. *Physics for Scientists and Engineers with Modern Physics: A Strategic Approach With Access Code*. Prentice Hall, 2012. 64

[218] B. Efron and R. J. Tibshirani. *An Introduction to the Bootstrap*. Chapman & Hall/CRC Monographs on Statistics & Applied Probability. Taylor & Francis, 1994. DOI: 10.1007/978-1-4899-4541-9. 74, 102

[219] A. L. Lloyd. The coupled logistic map: a simple model for the effects of spatial het-erogeneity on population dynamics. *Journal of Theoretical Biology*, 173(3):217–230, 1995. DOI: 10.1006/jtbi.1995.0058. 75

[220] K. Bache and M. Lichman. UCI machine learning repository, 2013. 88

[221] H. Ma and C. Han. Selection of embedding dimension and delay time in phase space reconstruction. *Frontiers of Electrical and Electronic Engineering in China*, 1(1):111–114, 2006. DOI: 10.1007/s11460-005-0023-7. 90

[222] D. Kugiumtzis. State space reconstruction parameters in the analysis of chaotic time series—the role of the time window length. *Physica D: Nonlinear Phenomena*, 95(1):13–28, 1996. DOI: 10.1016/0167-2789(96)00054-1. 90

[223] M. Sugiura and T. Kamei. *IAGA Bulletin N 40*. International Association of Geomagnetism and Aeronomy, 1991. 93

[224] M. A. Hapgood. Space physics coordinate transformations: A user guide. *Planetary and Space Science*, 40(5):711–717, 1992. DOI: 10.1016/0032-0633(92)90012-d. 93

[225] W. D. Gonzalez, J. A. Joselyn, Y. Kamide, H. W. Kroehl, G. Rostoker, B. T. Tsurutani, and V. M. Vasyliunas. What is a geomagnetic storm? *Journal of Geophysical Research: Space Physics*, 99(A4):5771–5792, 1994. DOI: 10.1029/93ja02867. 93

[226] R. K. Burton, R. L. McPherron, and C. T. Russell. The terrestrial magnetosphere—a half-wave rectifier of the interplanetary electric field. *Science*, 189(4204):717–718, 1975. DOI: 10.1126/science.189.4204.717. 93

[227] J. W. Dungey. Interplanetary magnetic field and the auroral zones. *Physical Review Letters*, 6(2):47, 1961. DOI: 10.1103/physrevlett.6.47. 93

[228] H. Tong. *Non-linear Time Series: A Dynamical System Approach*. Dynamical System Approach. Clarendon Press, 1993. 112

Author's Biography

JAMES M. MCCRACKEN

James M. McCracken received his B.S. in Physics and B.S. in Astrophysics from the Florida Institute of Technology in 2004, his M.S. from the University of Central Florida in 2006, and his Ph.D. in Physics from George Mason University in 2015. He currently lives and works in the Washington, D.C., metro area.